War Beyond the Battlefield

In an effort to make sense of war *beyond* the battlefield in studying the wars that were captured under the rubric of the "War on Terror", this book seeks to explore the complex spatial relationships between war and the spaces that one is not used to thinking of as the battlefield. It focuses on the conflicts that still animate the spaces and places where violence has been launched and that the war has not left untouched. In focusing on war *beyond* the battlefield, it is not that the battlefield as the place where war is waged has gone in smoke or has borne out of importance, it is rather the case that the battlefield has been dis-placed, re-designed, re-shaped and re-thought through new spatializing practices of warfare. These new *spaces of war* – new in the sense that they are not traditionally thought of as spaces where war takes place or is brought to – are television screens, cellular phones and bandwidth, George W. Bush's ranch in Crawford, Texas, videogames, popular culture sites, news media, blogs, and so on. These spaces of war beyond the battlefield are crucial to understanding what goes on the battlefield, in Iraq, Afghanistan, or in other fronts of the War on Terror (such as the homeland) – to understand how terror has globally been waged beyond the battlefield.

This book was originally published as a special issue of *Geopolitics*.

David Grondin is Associate Professor at the School of Political Studies of the University of Ottawa, Canada. His current research deals with the US preparation of and for the future of war and draws on the crucial links between sci-fi and cinema in imagining the transformation of war.

War Beyond the Battlefield

Edited by
David Grondin

Routledge
Taylor & Francis Group

LONDON AND NEW YORK

First published 2012
by Routledge
2 Park Square, Milton Park, Abingdon, Oxon, OX14 4RN

Simultaneously published in the USA and Canada
by Routledge
711 Third Avenue, New York, NY 10017

Routledge is an imprint of the Taylor & Francis Group, an informa business

This book is a reproduction of *Geopolitics*, vol. 16, issue 2. The Publisher requests to those authors who may be citing this book to state, also, the bibliographical details of the special issue on which the book was based.

Trademark notice: Product or corporate names may be trademarks or registered trademarks, and are used only for identification and explanation without intent to infringe.

British Library Cataloguing in Publication Data
A catalogue record for this book is available from the British Library

ISBN13: 978-0-415-52368-4

Typeset in Garamond
by Taylor & Francis Books

Publisher's Note
The publisher would like to make readers aware that the chapters in this book may be referred to as articles as they are identical to the articles published in the special issue. The publisher accepts responsibility for any inconsistencies that may have arisen in the course of preparing this volume for print.

Printed and bound in Great Britain by the MPG Books Group

Contents

The Other Spaces of War: War beyond the Battlefield in the War on Terror

DAVID GRONDIN

School of Political Studies, University of Ottawa, Ontario, Canada

Politics is the continuation of war by other means. Politics, in other words, sanctions and reproduces the disequilibrium of forces manifested in war.

—Michel Foucault, *Sécurité, Territoire, Population*

In his first days as President, Obama declared the 'end' to the 'War on Terror' – or so it was said in the media. If, in so doing, Obama was trying to get Americans to come to terms with the War on Terror as *past*, to refer to it in the past, it is granted that it is not an easy process and that it will meet reluctance, by Republicans especially. If Khrushchev had to wait Stalin's death before purging the Soviet Union from Stalinism, Obama cannot wait till George W. Bush is dead – and he has not done so, as his inaugural address made clear – nor can he sit idly by with regard to what he and many Americans see as the infamous Bush era plagued by the excesses of the War on Terror (e.g., torture, Guantanamo, Abu Ghraib, states and spaces of exception, absolute executive powers, etc.). Hence, as Germans learned after World War II, dealing with an ugly past is a daunting challenge and, in reference to their own dealing with their Nazi era and the Holocaust, they even created a word for it: '*Vergangenheitbewältigung*', which literally means 'Past [*Vergangenheit*]/ Coming to terms with [*Bewältigung*]' and which is better translated as 'trying to or struggling to come to terms with the past.' To enter in such process, while campaigning in 2008, Obama said he would do everything in his power to see the US '[get] out of Iraq and onto the right battlefield in Afghanistan and Pakistan.'[1] This was his first attempt at changing the course of US war-making enterprises in the War on Terror. Ever since he has been in power, he has set himself to get the US to abandon this 'metaphorical war' and at the end of March 2009, his administration decided to substitute the label 'Global War on Terror' with 'Overseas

Contingency Operations' to refer to the ongoing wars and military operations in Iraq and Afghanistan. Again, any attempt at coming to terms with the past, especially a recent one, will face resistance and this one makes no exception, as the War on Terror still rages, discursively and institutionally at the very least, albeit in varying degrees.

Our intentions here are less grandiose and do not seek to offer prognoses on whether the 'War on Terror' has really ended and what is to be expected by the Obama administration's taking over what was associated (in terms of programmes, practices, and discourses) with the idea of the War on Terror. Instead, more humbly, in trying to assess the nature of the War on Terror – rendered as an open-ending venture *and* as a 'war' by the George W. Bush administration – we aim to make sense of understanding war *beyond* the battlefield in the 'new' strategic environment set forth by the matrix that became the War on Terror. War is a social phenomenon that generally happens and that takes place on a battlefield. But the battlefield made possible by the global information technology rests on a different register of spatialities and temporalities. In a revolutionary way, one such space is a mobile and global one, one that has also transformed the nature of warfare itself, as Paul Virilio rightfully pointed out a while ago:

> *The battlefield is at first local, then it becomes worldwide and finally global*; which is to say expanded to the level of orbit with the invention of video and reconnaissance satellites. Thus we have a development of the battlefield corresponding to the development of the field of perception made possible by technical advancements, successively through the technologies of geometrical optics: that of the telescope, of wave-optics, of electro-optics; that of the electromagnetic transmission of a signal in video; and, of course, computer graphics, that is to say the new multimedia. Henceforth *the battlefield is global. It is no longer worldwide* [*mondialisée*] *in the sense of the First or Second World Wars. It is global in the sense of the planet.* For every war implicates the rotundity of the earth, the sphere, the geosphere.[2]

The focus is therein given on the *impacts* of war more than on the *conduct* of war by its main protagonists on the battlefield.

This special issue seeks to interrogate and discuss the new spaces, theatres, and realities of war beyond the battlefield that the politics of war involved in the War on Terror has enacted. Through a Foucauldian understanding of war as politics, authors in this special issue see the War on Terror as a spatialised framework where the boundaries of war and politics collide. Accordingly, this special issue on the 'war' on terror is not your traditional strategic analysis of war nor is it a neoclassical geopolitical analysis; it is actually less interested in what happens or has happened on the battlefield – traditionally represented as theatre of operations where war is waged – than it is on assessing what happens when the battlefield is morphed into new spaces and when it is displaced, when it becomes a semiotic space for making sense of the politics of war and of war as politics. In focusing on war *beyond* the battlefield, it is not that the battlefield as the place where war is waged has gone in smoke or has born out of importance, it is rather the case that the battlefield has been dis-placed, re-designed, re-shaped and rethought through new spatialising practices of warfare. These new *spaces of war* – new in the sense that they are not traditionally thought of as spaces where war takes place or is brought to – are television screens, cellular phones and bandwidth, George W. Bush's ranch in Crawford, Texas, videogames, popular culture sites, news media, blogs, and so on. These spaces of war beyond the battlefield are crucial to understanding what goes on the battlefield, in Iraq, Afghanistan, or in other fronts of the War on Terror – to understand how terror has globally been waged beyond the battlefield.[3]

This collective effort builds on the footsteps of recent work in critical geopolitics, notably two recent collections, Rachel Pain and Susan Smith's *Fear: Critical Geopolitics and Everyday Life*[4] and Klaus Dodds and Alan Ingram's *Spaces of Security and Insecurity: Geographies of the War on Terror*[5]. In Pain and Smith's edited volume, what is rightly underscored is how the very notion of the War on Terror does not bring in a 'new' fear, but rather how emotional landscapes of everyday life are construed through fear so as to appear as breaking through with the past. As they assert, with September 11, 2001, 'it was not the dawn of a *new* era, either for the nexus of international relations and everyday life or for the spatial politics of fear. The historicized and ground-truthed accounts of fear which we present in this collection make the continuities clear'[6] (original emphasis). In Klaus Dodds and Alan Ingram's collection, the spaces of security and insecurity that the War on Terror has contributed in erecting are situated on a critical geopolitical plane. Like this special issue, its interdisciplinary focus bridges the fields of critical security studies, critical geopolitics, and critical geography, as well as critical and feminist International Relations in order to expose how security has been conceived of as governmentality of the homeland, of borders, of citizenship, and the numerous spaces gazed by sovereign and biopolitical power in the War on Terror. What it especially brings to the fore is how specific geographies and practices of securitisation have woven geopolitics and security together 'at specific sites and through scalar processes, and particularly in relation to mobile or transnational populations or entities.'[7]

In thinking war beyond the battlefield, we are also led to highlight the explosion of popular cultural analysis in both IR and critical geopolitics, and more prosaically on the representations of war and global politics.[8] Jason Dittmer's latest monograph, *Popular Culture, Geopolitics, and Identity*[9] then reveals as a worthy addition to the literature on popular geopolitics and it represents a successful attempt at bringing critical geopolitics further along in its effort to study popular culture to account for geopolitical issues. As a work of popular geopolitics, it does steer critical geopolitics towards a better understanding of everyday life practices and its attempt to better connect both the affect that comes with consuming popular culture and the critical attitude that a critical geopolitical stance commends is successful. In this respect, the term he has coined of 'performative consumption' pictures this very well. It

> describes the double-ness of engaging with popular media – not only are you consuming, for instance, a war movie and constructing meaning from the dialogue and images, but you are also performing an identity when you do so. For instance, you could be watching the movie with a sense of gung ho excitement, or you could be watching it ironically, laughing at the one-liners that are meant to be inspirational. Thus, engaging popular media is not just passive, as many critics of TV or pop music would have it – it's also active in that you shape the ways in which you are oriented to that popular medium.[10]

In a like-minded spirit, whereas Dittmer gets us to rethink the War on Terror through popular culture, this special issue is interested in deciphering the multiple realities that the very efforts of thinking 'war beyond the battlefield' make visible, the pluralist ways that war gets represented, framed, and understood, and the contexts and consequences of the actions and decisions that policy-oriented analyses either take for granted or rule out as not being worth investigating further. While the War on Terror and the ensuing War in Iraq set the context, subtext, and text of the analyses comprising this special issue, the authors are here engaged in recreating the contexts of certain actions and decisions, in evoking imagined and real consequences of this global war as new normalcy in contemporary global politics, and in making possible a critique of the War on Terror and the systems of representations that made its occurrence possible. We have all witnessed and lived through the War on Terror and watched the on-screen unfolding of the War in Iraq and Afghanistan, and, yet, there is still so much about these events and their effects on the world we have already forgotten or simply are unaware (like how the liberal way of war and the liberal way of rule work as corollary modalities of biopower in our current global predicament).[11] Somewhat akin to what Fraser Macdonald, Rachel Hughes, and Klaus Dodds, the editors of the recent study *Observant States: Geopolitics and Visual Culture*, which introduces visual studies to critical geopolitics, aimed to achieve in their prescription against an aestheticised view of September 11 that would befall from '"the visual grammar" of display [which] . . . mobilize political action through the affect of the visual', in its treatment of images and events this special issue has deliberately avoided treating the conventional analysis of war on the battlefield or its visuality of the battlefield and rather sought to make war beyond the battlefield visible.[12]

THINKING WAR BEYOND THE BATTLEFIELD
IN THE WAR ON TERROR

The United States of America led by the George W. Bush administration was not an innocent bystander in the September 11, 2001 attacks – as this would be tantamount to *not* taking notice of the strong resentment of US hegemony abroad, especially in the Middle East countries and in the Muslim regions prior to these events. The connection of two seemingly different military operations in the guise of the US military operations *Enduring Freedom* in Afghanistan and *Iraqi Freedom* in Iraq as part of the same ongoing global military war effort aimed at countering and defeating a global *Jihad* bewildered many and generated questions about the new governmentalities produced by the War on Terror. With the US branding of the War on Terror as the war *for* civilisation, as *the* 'just war' humanity should be ready to fight, it then came as no surprise that the extension of the War on Terror in Iraq provided a new breeding ground for Al Qaeda and Islamic terrorists to fuel their rage against American interests and what they saw as the global projection of Western hegemonic power and ideals.

The War on Terror has so far required lots of sacrifices on all sides and its extension in Iraq and Afghanistan has meant even more – and the US state leaders' deceit was certainly seen as being part of the problem more than the solution. Now that what became the 'war effort' in Iraq has become somewhat of a 'successful' surge, allowing the Obama administration to opt for an exit strategy in 2010 in order to shift its attention to Afghanistan (and Pakistan) instead, it may yet be time to try to assess the nature of the War on Terror as it unfolded *beyond* the battlefield in the wake of September 11, 2001.

GOVERNING WAR AS POLITICS: THE LANGUAGE OF WAR

As critical geopolitics scholarship taught us, geopolitical and strategic 'realities' are produced by language, and in its study of geopolitics, 'critical geopolitics encompassed various ways of unpacking the geographical assumptions in politics, asking how the cartographic imagination of here and there, inside and outside, them and us, states, blocs, zones, regions, or other geographical specifications, worked to both facilitate some political possibilities and actions and exclude and silence others.'[13] Hence, to study geopolitics, one studies how the world is thought, said, written and how, once inserted into narratives, such narratives constitute discourses that (re)produce 'reality' – where discourses are ways of producing something from the 'real world' as real, as identifiable, knowable, and meaningful.[14] From the get-go, it is sure that the label itself, 'War on Terror', is deeply problematic, as were the expression 'Cold War' or 'equilibrium of terror' used to describe the geopolitical confrontation between the USA and USSR and the nuclear logic of annihilation that rested on mutual assured destruction between them. In and of itself, a signifier such as 'War on Terror' is woven into the domain of affect, as it speaks of a powerful emotion – terror – that represents a heightened fear, induced so severely that one subject has come to wish to counter, master, conquer, and defeat it by declaring war on it, the United States of America of George W. Bush.[15]

In many respects, there seems to be an increasingly evident disconnection today between what scholars, politicians, and citizens have to say about war. Today's War on Terror may not be a war in a traditional sense, but the states' resources and war-making apparatus are involved as if it were fighting a 'real' war. It is exactly what President Bush seemed to mean in the aftermath of September 11, 2001 when he said that the American response would not be a war in the way we were used to think about wars, even though it would include using military force.[16] The crucial issue to understand is that ordinary people think of war as a last resort measure involving military means that are only needed to solve a political violence inflicted on their nation/state. However, since the 1980s, policy makers and several scholars have used the conceptual language of war to address social problems of violence. Seen in this scheme are the government's 'war on poverty' and 'war on drugs', two metaphors that abhor how problematic the issue of agency and political violence are constructed. Governmentalised wars on poverty and drugs reflect how the intent is to 'defeat' a problem as if it were cast as an 'enemy'. The phrasing of war to engage terrorism also indicates 'a certain power structure in the relations among discoursing subjects (as well as silenced subjects).'[17] As James Der Derian captured it, 'The war on terror is not new but part of a permanent state of war by which the sovereignty of the most powerful state is reconstituted through the naming of terrorist foe and anti-terrorist friend.'[18]

In various ways, the Bush administration's war on terrorism in Afghanistan, Iraq, and on its own homeland reminded us how deeply entrenched is the militarised thinking and the use of military forces to solve problems that appeared more as social and political problems.[19] In effect, the war on terrorism metaphor sprung into a galvanised yet renewed effort to wage war when spaces for alternative or critical thinking should have been made possible rather than being marginalised, ignored, or deemed unpatriotic. It is important to understand how the War on Terror has rejuvenated the whole strategic enterprise of US warfighting and military preparedness. As Mikkel Rasmussen aptly points out, 'It is a cause for concern that much of Western expertise on security issues is simply dismissing the agenda of the 'war on terror' as a politically dangerous and intellectually unsustainable project . . . . it might also be the case, however, that far from being an anomaly, the 'war on terror' is an example of a new *practice of security*.'[20] From the offset, the War on Terror implies an engagement with the issue of risk, of technology, and governmentality for us to assess its perceived and hidden effects in our everyday contemporary life.[21]

THE WAR ON TERROR AS GOVERNMENTALITY

In trying to conceptually make sense of what the War on Terror actually entails, one has to unpack the new governmentalities at work – thus, address how the War on Terror acts as *governmentality*. As Foucault taught us, governmentality comprises 'all the institutions, processes, analyses and reflections, calculations and tactics that allow for the exercise of this specific form of power.'[22] What is more, through governmentality, the logic of sovereignty is gradually subsumed as power enacted by the state, as governmentality permits to go beyond the state's sovereign power, achieved from above, and to exercise power from a more decentralised mode, as well as to foster more efficient forms of management.[23] This Foucauldian understanding of dispersed power enables us to address the War on Terror as a matrix and in a more systematic way as a producer of new forms of government of our everyday life. Foucault's work on liberal governmentality, security *dispositifs*, and war provide the necessary theoretical background to expose, explore, understand, and critique how new governmentalities geared to secure the homeland actually work and manage the global order. It allows us to examine how it transforms our beings and our ways of relating to space and place, to the world, to the local and the global – to think security 'at home'.[24] We thence propose to formulate a way of understanding the War on Terror that goes beyond the war on terrorism and the War in Iraq *even though* it remains bound to them, an understanding that also comprises homeland security activities.[25]

From this account, we may stipulate that the War on Terror is a complex matrix of discourses and institutions configured by practices that are not exclusively military that led to new governmentalities regimenting several types of practices that involve as much policing work as routinised security and military activities that go beyond the turf associated with an exceptional context of war.[26] As a result, the War on Terror is as much a war that invokes a battlefield – as it involves military combat operations and strategies in Afghanistan and in Iraq[27] – as it is a war defined through a Foucauldian understanding of 'politics as the continuation of war by other means' (an inversion of Clausewitz's famous dictum of war as the continuation of politics). War is also an activity that is long planned, rationalised, and intentionally circumscribed. War will always break from its 'crafters', as its unfolding will redirect the political aims that initially issued its implementation in the first place. This one – the War on Terror – has certainly escaped the Bush administration's original intent. Moreover, the very unpredictable nature of war means that one should factor in this unpredictable character into any supposedly rational decision-making process which elects war as a policy tool. Accordingly, focusing on the governmentality of the War on Terror and the unexpected side effects of this war entail many reflections on the importance of understanding war 'beyond the battlefield', where one emphasises the experiences and effects of the War on Terror *as war* on societies and individuals.[28] In this context, war needs *not necessarily* be what many see to be the 'classical' view, that is, the deliberate application of organised violence by one group of people and where 'deliberate' and 'organised' presume a planning, rehearsal or simulation, and an orchestration of violence to inflict harm on a target.

War has exploded through information technology and the War on Terror is a prime example showcasing the multifaceted ways that war may govern and transform our lives, especially when it is being cast as the 'first Internet war.'[29] This 'iWar', as dubbed by Mark Andrejevic for its interactivity, gave a role to the Americans citizens who would also participate in its self-governance, especially through its micro- and macroscalar deployment on the Internet and other spaces of societal and political life. Surprisingly, this interactive facet of the War on Terror highlighting its governmentality as partly being the governmentality of the self was best rendered by none other than the former Republican senator from Pennsylvania, Rick Santorum. He summarised the War on Terror as being 'a truly modern war – a war fought not just on the battlefield, but on the Internet, a war decided less by armies and warplanes than by individuals making individual choices,'[30] to which Andrejevic added that it was a 'war to be waged by individuals making the "right" choices.'[31] Such account of the War on Terror feeds in the unpredictable aspect of any war and imprints its governmentality with even greater complexity. By all means, this is inescapable, as Chris Hables Gray tells us: 'War, as a very complex and volatile system, cannot be controlled, it cannot be managed and it cannot be predicted. This is as true of real bloody war as it is of "virtual cyber war", as if any war that didn't damage and destroy bodies could really be called a war.'[32] The very logic of war is that of irreconcilable contradictions between adversaries; the only way to get rid of the possibility of war would be to break free from an attitude of taking risks. But can humanity really be able to regulate war and impose a legal codification of war that all would adhere to? If, in Clausewitz's view, war is always inescapably a *political* instrument – and hence the inverted view of politics as the continuation of war – war must always remain an open possibility, even if an unlikely one. Henceforth, war more than ever is 'political and not technological. . . . it still comes down to what is done with messy bodies.'[33] This is true of the War on Terror as well, even though how war connects these bodies into narratives is still up to analysts and scholars to decipher. This special issue is one attempt at inscribing these narratives in other places and spaces of war, beyond the battlefield.

SPATIALISING THE WAR ON TERROR AS TRANSVERSAL PHENOMENON

This special issue is therefore an exploration of the complex spatial relationships between war and the spaces where one is not used to thinking of as 'the battlefield.' The War on Terror has led many to reconceptualise the ways armed forces were geared to wage war and were conducting military operations. What is the influence of the representations of such evolutions in media? In filmic representations? In the context of the War on Terror? In invoking war beyond the battlefield, this special issue focuses on the conflicts that still animate the spaces and places where violence has been launched and that the war has not left untouched. The idea is to bring about a space were the representations of war beyond the battlefield can be revisited and contested, where the global spatial politics of war which still unequally affects us all can be engaged in places and spaces where they are usually not expected nor welcome, where the materiality of conflict and war can be revealed *beyond the battlefield*. This is obviously done to produce alternative ways to think the geographies and spaces of the War on Terror, as well as the practices of warfare and their impacts that make up the War on Terror. In this very sense, the War on Terror offers itself as an excellent case in point for what Roland Bleiker calls a 'transversal phenomenon', where one construes new spaces by looking at several simultaneous political and social transformations in a stance that 'not only transgresses national boundaries, but also questions the spatial logic through which these boundaries have come to constitute and frame the conduct of international relations.'[34] David Campbell argues that our current global everyday life is best represented 'as a series of transversal struggles rather than as a complex of inter-national, multi-national or transnational relations.'[35] As Campbell explains:

> Likewise, neither is 'everyday life' a synonym for the local level, for in it global interconnections, local resistances, transterritorial flows, state politics, regional dilemmas, identity formations, and so on are always already present. **Everyday life is thus *a transversal* site of contestations rather than a fixed level of analysis**. It is transversal because it 'cannot be reconciled to a Cartesian interpretation of space.' And it is transversal because the conflicts manifested there not only transverse all boundaries; *they are about those boundaries*, their erasure or inscription, and the identity formations to which they give rise.[36]

In this special issue, authors ask limited questions in a field, searching through events, space and time to make sense and give intelligibility to events transversally inserted in a problematique: *experiencing war beyond the battlefield in the War on Terror*. The analyses gathered for this collection of essays are more than literary or prosaic exercises; they are *critical interventions* that ask questions and offer some elements of answers on events that are part of the War on Terror, that *make* the War on Terror. As such, any utterance of this 'War', as lived on the battlefield and beyond, may refer to the same war, but it is a war that will differ according to the subjects living and writing (on) it (e.g., those who fought it and those who suffered from it, those who were drawn into it, and those who analyse it, etc.). So no matter what the conflict is, once played out, each war has its own stories and meanings – waiting to be told, unveiled, or reinterpreted. In a way, this special issue pays heed to sociologist Catherine Lutz who, in her effort to explore 'how America's military has affected daily life in this country', argued we should not 'imagine the costs of war as exacted only on the battlefield and the bodies of soldiers . . ., as civilians are now the vast proportion of war's clotted red harvest', but strive to make visible the 'blurred boundaries of the civilian and military worlds', in other words, we should decipher what happens beyond the battlefield.[37]

COMING TO TERMS WITH THE PAST OF SEPTEMBER 11, 2001

Almost ten years after September 11, 2001, there is no denying that the War on Terror *haunts*[38] us; we have all witnessed and lived through the War on Terror and the War in Iraq. Given that war as one has imagined it is as much a war as an actualised war,[39] it becomes crucial to reflect on its imagination in the same way as on its actualisation. In other words, looking through the fictions of war, the simulation of war, the virtualisation, or the visualisations of war can teach us as much as what looking at the reality of war can; this is why, for instance, movies and documentaries – as cultural productions – offer good sites of political analyses of the war itself. In effect, through the analytical process, one will be able to sift through the symbolic and imagined representations of worldly situations that are unwittingly tied to the War on Terror and reflect on, say, an agent's intentions, goals, beliefs, and certitudes as if they were real. As French historian Marc Ferro aptly remarks: 'This goes with the remembrance of war as war itself: it varies according to the memories; those of the political leaders and chiefs are not necessarily those of the warfighters nor that of those who stayed behind – the memory of the winners is also not that of the losers. There exist zones of memory that survive, others that die, others that remerge in the most unexpected moments.'[40]

Acknowledging the always contestable character of the constructed narratives about the past, present, and future in their discursive appropriation, these (re)interpretations of the War on Terror – in the experiences of home-land security, of the War in Iraq, or of the political act of dissenting from the war, among others – are open invites to reflection, meditation and reflex-ive assertions aimed to provoke and generate new questions and discussion about the real that the War on Terror encapsulates, whether one understands it *as* event, war, politics, matrix, era, or global governmentality.[41] Indeed, if 'the war that has not happened yet is as much history as history,'[42] then surely the one we are currently living in the past and in the present *is*, too.

Adopting Foucault's 'interpretive analytic of our current situation', his 'ironic stance toward one's present situation', Hubert Dreyfus and Paul Rabinow invite us to his 'unique combination of genealogy and archaeology that enables him to go beyond theory and yet to take problems seriously.'[43] Following the months of September 11, 2001, we might have thought that it was not yet time to reflect on and dissent from the US-led global war on terror(ism). We might have felt unease discussing rationally then, we may have believed that American insecurity and the Bush administration cry for revenge was as 'normal' a reaction as we would have expected from another state/people. As we have (finally, some might say) turned our back on the Bush administration and as the Obama administration has struggled in turn-ing hers on the War on Terror, it is more than time, in an era plagued by and still trapped in the War on Terror, to critically engage with our past, present, and future as affected by the global predicament that is the War on Terror.

INTRODUCING THE COLLECTION OF PAPERS: WAR BEYOND THE BATTLEFIELD IN THREE ACTS

The collection of papers is organised in three successive acts. The special issue first emphasises the governmentalised experience of the War on Terror as it relates particularly to the whole enterprise of what securing the home-land entails. It then represents the visual experience of war fought on and beyond the battlefield and undertakes to make us rethink how the experi-ence of the reality of war may affect us through its video gaming simulation and through the contested mediascape produced by the visual technologies of real clips of the War in Iraq. Finally, it focuses on the bodily experience of war as politics, exploring how war may create possibilities for the politi-cal and how war is always made on (certain) bodies and differently affects societies and individuals.

Act I—The Governmentalised Experience of the War on Terror: Securing the Homeland

In an effort to criticise familiar logics of geopolitics where centralised state powers try to establish a strategic and spatialised control over territories, populations, and resources, the first section of the special issue seeks to render unfamiliar the way the War on Terror has been understood and received and to acknowledge how it has led to new forms of government and security, and how it has produced new global governmentalities. Indeed, the very thinking about collective and global security has underdone radical changes, and through the US leadership in this global War on Terror, new practices that wish to better secure homelands have emerged under the heading of 'homeland security' (such as the implementation of the Department of Homeland Security by the United States).[44] For this very reason, it becomes imperative to take notice of how state managers of security and police have adapted liberal governmentality to the newly created spaces and discourses of 'homeland security' and, thus, to look closely and critically how the War on Terror has been waged at home and globally.

On that account, the first two contributions, written respectively by John Morrissey and Miguel de Larrinaga, look at the War on Terror in terms of its strategy planning and of its actualisation abroad in the Middle East *and* globally. In considering the central role this 'region' and its spatialisation in an all-encompassing military strategy have played in the occurring of a 'Global War on Terror' – as it is understood that this *global-isation* of the war has partly been constructed by US strategic planners – Morrissey decides to pay attention to the United States Central Command (US CENTCOM), whose 'Area of Responsibility' in terms of US 'total world surveillance' encompasses Afghanistan, Iraq and the greater Middle East region.[45] Whereas Morrissey looks at the long-planned preparation of what became known as the War on Terror, de Larrinaga opts for a transversal look at the War on Terror, the borders of which were made global by the alleged nature of the threat of Islamic terrorist networks and by the response to September 11, 2001 made by most state leaders, and by the US state especially. The subsequent piece, owed to Jeffrey Bussolini, is more directly concerned with this (r)evolutionary aspect of the War on Terror, the focus on homeland security. Using Giorgio Agamben and Michel Foucault's work, Bussolini addresses a new space of war beyond the battlefield in looking at production sites of national security in the United States in the context of nuclear weapons and the biopoliticising effect their advent had on a totally mobilised war society during the cold war. As a result, through this framing of homeland security, security is *really* 'coming home.'[46]

Preparing the 'Juridical' war

As he invites us to unpack the global pretention of what became known as the US 'Global War on Terror' through an explanation informed by Foucault's *Security, Territory, and Population* Collège de France's lectures on the governmentalisation of security, Morrissey focuses on a biopolitical critique of the War on Terror storyline to bring his rendition of war beyond the battlefield to look at 'a "juridical" form of warfare, or lawfare, that sees US troops as "technical-biopolitical" objects of management whose "operational capabilities" on the ground must be legally enabled'.[47] This uncommon biopolitical focus turns away from the civilian populations to target the American troops in order to assess more precisely 'how the US military actively seeks to legally facilitate both the 'circulation' and 'conduct' of a target population'.[48] As he discusses the 'forward deployment beyond the battlefield', Morrissey digs deeper in exposing how US military strategy has been strategically deploying a decisive legal maneuver in its effort to maintain a 'forward presence' in the one strategic region it had first targeted, the Middle East: what he calls 'the securing of "Status of Forces Agreements"'.

As this topical interest in this key player in the advancement and management of the 'Global War on Terror' is highly dependent on the military geographical knowledge of the Middle East and the geo-strategic considerations of US national interests (in oil resources especially), it reveals all the more crucial to realise that in looking at the projection of American military and geopolitical power in the Middle East by US CENTCOM, a dialogic and reflexive space is produced to interrogate how the conditions of possibility were set in motion for fighting a 'War on Terror' before the September 11, 2001 events had even occurred. Morrissey can therefore undertake a much-needed partial genealogy of the present War on Terror that recasts the Middle East as one space of exception gone global in the aftermath of September 11, 2001.

Along this line, using Giorgio Agamben's notion of the 'state of exception,'[49] Morrissey makes particular use of CENTCOM's published strategy paper on the Middle East, *Shaping the Central Region for the 21st Century*, to interrogate the specific ways in which US grand strategy invokes and perpetuates a series of 'states of exception' in both the imagining and functioning of US geo-strategic interests across the region. Indeed, CENTCOM's geopolitical and geoeconomic mission partakes in the creation of a global prison system establishing 'zones of exceptions'[50] – where law is strategically suspended for some by US security strategists – and is part and parcel of the 'US military's "grand strategy of security" in the war on terror' which 'includes a broad spectrum of tactics and technologies of security, including juridical techniques . . . relentlessly justified by a power/knowledge assemblage in Washington that has successfully scripted a neoliberal political economy argument for its global forward presence.'[51] This 'grand strategy of security' alluded to by Morrissey is the very turf of what I have elsewhere termed 'US national security governmentality' and it relates to these state and non-state elites working along the parameters of national security, whether they are *in* government or acting *as* governmental 'non-state scribes' for the state and who project US national security discourse onto the 'global' as a frame of operation for governance.[52]

WAGING A GLOBAL WAR

Miguel de Larrinaga's piece is a logical follow-up on Morrissey's piece. War is here seen as productive. Sharing its title with a Beatles' song in which John Lennon wove together several newspaper stories from a particular day to create 'A Day in the Life', de Larrinaga focuses on a specific day's events to examine transformations in the current world (dis)order: November 20th, 2003. He uses this day as a barometer – as a tomogram, in de Larrinaga's Foucauldian wording – to examine global governmentality and the current predicament of the representation of global order. On this day, several events occurred, including the bomb attack on the British consulate and the HSBC bank headquarters in Istanbul; a presidential visit by George W. Bush to London accompanied by antiwar protests; suicide bombings in Kirkuk and Ramadi; an evacuation of staff from the White House due to a 'blip' on a radar screen rather than a plane; anti-FTAA protests and clashes with police in Miami, and so on.

De Larrinaga frames his analysis within the broader context of a questioning of the 'eventness' of the event, of the role of agency and of the importance of territoriality in contemporary world politics, as well as the process and significance of dating in representing global order. In so doing, he attempts to highlight the tensions between global politics understood and articulated from a sovereign optic and an understanding of global politics as a site of transversal struggles in a world where '9–11' and the War on Terror provide the fundamental markers of current representations of global order. This piece presents well how this governmentalised war experience has global ramifications and has produced new spaces of government of the global as well as the limits of a traditional legalised and static understanding of sovereign power.

WAGING WARFARE AT HOME

Using the case of the United States in deploying a homeland security strategy geared to defeat terrorism at home through Foucault's *Il faut défendre la société*, Jeffrey Bussolini's piece presents evidence, from the site of Los Alamos, New Mexico, that the subjecting of the entire population to the absolute risk of death (in atomic weapons) both reveals a new aspect of sovereignty and is a crucial precursor to the biopolitical administration and power over life. In the account canvassed by Bussolini, it is the face of the securitised and militarised city that is brought to bear with the 'invention' of Los Alamos as the 'nuclear borderlands',[53] 'as a model town . . . [that] emerged as a vision hewn from the Atomic Age and a place for families to adapt to the challenges of the Cold War', where 'the mystique of the Atomic City combined with the vital national security role of the laboratory to attract the nation's attention'.[54]

He looks at how, since September 11, 2001, through a language of homeland security, the US state has focused on further developing defensive strategies to protect cities from the threat of political violence, emphasising especially the social effects of securitisation/militarisation on changes in the nuclear state and a biopolitical society. Concerned with the social effects of securitisation, i.e., the wider social-structural and social psychological effects of a totally mobilised war society, Bussolini illustrates how the US national security state, through the development of nuclear technology internationally, attempts to better secure its homeland by being prepared to wage war at home. In that sense, Bussolini provides us with a biopolitical history of the nuclear age through the case of the Atomic City that is Los Alamos, where 'the domestic space itself became coterminous with the battlefield in the nuclear age' with the nightmarish prospect of nuclear annihilation and mutual assured destruction. In other words, Bussolini is able to connect the Cold War to the War on Terror through the aegis of homeland security.

Act II—The Visual Experience of War Fought on and beyond the Battlefields: Rethinking (the Experience of) War

In this second portion of the special issue, the essays take issue with the visual, actual, virtual and real experiences of war on the battlefield and beyond, as well as their imagination. In the spirit of thinking through war beyond the battlefield and in other spaces of war, Mark Salter addresses the issue of the virtualisation of war in the spaces of everyday life, questioning how war videogames and war simulations make the bodies disappear and transform the warrior experience – where war is really *yet* virtually fought beyond the battlefield. With Benjamin Muller and John Measor's piece, we turn to *seeing* the actual experience of war-fighting in the Iraqi war through a mediated portrayal of the violence on the battlefield and an analysis of the reception of its visual images beyond the battlefield. Though media have long been crucial on and off the battlefield,[55] they provide us with a new media landscape ('mediascape') that opens our eyes to the diversity of fictionalised and more intimate accounts of the war experience amongst those opposing, resisting, and suffering the US occupation, as well as to the clash between official media representations of US armed forces actions and their own personal attempts at 'telling *their* story'. Both contributions incidentally and powerfully bring out this other 'affective' space of war that both profoundly shapes how war is waged and mediated on and beyond the battlefield.

SIMULATING WAR/PLAYING WAR

In his piece, Mark Salter takes on virtual war as it becomes a new space of war – as war-gaming and war simulation are essential to the war machine, to recruiting, to training, and to the very conduct of warfare – while also being part of a growing entertainment industry.[56] Salter is here committed to undertaking a serious critical geopolitical reflection on the extent of the social and cultural processes of militarisation through video games – where video games are read as texts of critical geopolitics – and explores 'the construction and contestation of the popular international geographical imaginary'. Through an examination of Foucauldian schemas of 'politics as the continuation of war by other means' and a comparative analysis of *Diplomacy, Civilization, America's Army*, its accompanying 'Virtual Army Experience' (VAE), and *Grand Theft Auto IV: Liberty City*, he apprehends the visual experience of war as one of simulation, where the warrior is virtually present on the battlefield. His critical geopolitical analysis thus mulls how corporealism is affected by virtuality, inquiring in video games how representations of virtual spaces and bodies influence corporeality.

Scrutinising the new American geopolitical imaginary at play in the War on Terror to prepare the transformation of war beyond the battlefield, Salter argues that, from his analysis of *Grand Theft Auto IV*, 'New York City [became] a localized site of the global war on Terror', one that 'demonstrates the reorientation of the American military imaginary towards the domination of urban space and surveillance'.[57] His analysis closely follows Stephen Graham's work on military urbanism. In effect, it emphasises the urbanising of security, where 'circulations and spaces of the city are becoming the main 'battlespace' both at home and abroad'.[58] Produced for different spatialities and fields of intervention but nevertheless deployed on and beyond the battlefield – the urban space at home and abroad – this new 'military urbanism' gets security planners to gaze at the urban space of homeland security so that it gets controlled, monitored, and represented *as* that of a battlefield. As a result, American cities such as New York and Los Angeles are targeted by state military doctrines of urban warfare, in a power/assemblage of urban governance of public safety and policing activities linking police and security operations at home as part of the global war on terrorism. In this context, in an attempt to better secure the homeland by being prepared to wage war at home, cities are deemed protected from the threat of political violence through a wide range of defensive strategies rooted in military doctrine, thus militarising and policing even more American everyday life – as it is portrayed in the virtual city of *Grand Theft Auto IV*. The counterpart of the 'urban battlefield', the one that takes place abroad, ventures 'the wholesale destruction of urban civilian infrastructures as ways of bringing 'strategic paralysis' to urbanized adversaries'.[59] This 'urbanisation of the battlespace' which benefits from playing videogames explicitly aims to disrupt cities — and lifestyles — by targeting critical infrastructures and, consequently, *civilian* populations. This 'urban turn' in the 'Revolution in Military Affairs' profusely imagined and featured in urban warfare videogames also pays tribute to the RMA legacy of wishing to sanitise and aestheticise killing that we find recast through the discourse of the 'liberal way of war'.[60]

In line with Chris Hables Gray's take on postmodern war, Salter argues that insofar as war 'still comes down to what is done with messy bodies,'[61] the trick of war is nevertheless to make bodies disappear: 'The body politic of the state is transformed, the bodies of soldiers are obscured, and enemies are rendered into body parts.' This then leads him to argue that the virtualisation of war tightens this 'double-bind of the war-body politic' in the decorporealisation of the gamer-subject-citizen playing *America's Army* or *Grand Theft Auto IV*. Salter finally demonstrates how 'tensions between the governance, biopolitical and tactical war-fighting representations of the city' could be extremely productive for critical geopolitics.[62]

SEEING WAR (MEDIATED)

Benjamin Muller and John Measor's piece makes us wonder about our acceptance of reality and our need to see everything – *to be* on the battlefield through visual technologies; where *the visual technologies are the battlefield*. They discuss images of the War in Iraq taken by Iraqi insurgents and US armed forces personnel, images that are then relayed via the Internet in global and local social networks. Their analysis of these web-based media creations broaches a new media and emotional geography that stresses how speed and Web 2.0 technology – in the form of viral videos especially – have contributed to our understanding of how distance, places, and, emotions may connect those who consume these media productions. Through their account of the circulation of conflicting representations of identities, events, and 'lived experiences' wired through official and more informal media, they are able to depict how the conflict found its way in the digital mediated accounts of US military personnel and Iraqi insurgents. They illustrate how a 'battle in the wires' relying on visual technologies seeking to represent the battlefield becomes *the battlefield*. Notwithstanding the very problematic capture of the battlefield images, they show how images of the War in Iraq really have the potential to generate a political and social involvement from people who would otherwise not have been sensitive to these issues. In fact, they illustrate how this reveals as one good opportunity for a certain civil society to organise itself without the state's oversight. In a media context where mainstream America is being offered an 'authentic' portrayal of the spectacle of war on television screens because of the several embedded journalists accompanying US armed forces, these disturbing images and *actions* (of filming and circulating these images) may well bear an important effect on the lived experiences of everyone on and beyond the battlefield, and in a way that mainstream media never would nor could achieve.

As Muller and Measor acknowledge, representations of battles, warriors, and battlefields have always been ways to inform the wider citizenry about violent conflicts to which they were mere distant observers. If one's vision of a battlefield is never the same as another's, from the very moment that these images are *produced* by actual combatants themselves instead of 'observers' reporting what is happening, it can be assumed that the impact may be greater on the perception of the citizenry as the sources would seem to be more credible – as if these images were not filtered in any way, as if they were authentic and raw images, reality unveiled... Certainly, this raises the question of our relationship to technology, which cannot be ruled out as either good or bad, as Muller and Measor concur. It transforms war, and our understanding of *this* war, as it brings up questions such as: Do we need to see crude and raw visual images to be convinced of war's atrocities? What is the lasting effect of these images of the real on those who consume them? For concerned citizens who get used to seeing and receiving these images, does it help them build a more critical view on war or does it make them accept war as normalcy? For sure, presented as recruiting material in the case of Iraqi insurgents, it is pretty clear what is intended with these images. If it remains difficult for the US military to control these images and assess the damaging effects of these actions, at home and abroad (it is difficult here to escape the effects of the Abu Ghraib torture photos abroad and at home), it becomes interesting to inquire how the othering of Iraqis – that the clips taken by the US military personnel represent – has been made possible. We can hence acknowledge how warfare is so heavily primarily a mediated phenomenon, even when discussing war beyond the battlefield.

Act III—The Bodily Experience of War as Politics: War as Possibilities for the Political

The final section of the special issue emphasises the bodily experience of war as politics. These essays take us in more intimate spaces, in the public, private, personal, sociological and social spaces in which subjects such as citizens and scholars evolve. The objective is to explore the familiar, strange, and ambiguous feelings that war and its bodily experience interpellate and the insights that war beyond the battlefield generate. We are all unconsciously already participants in war through the lived experiences of war or its effects on human beings, as we are all transformed by the discursive experiences of war as lived: 1) by a sibling or a friend; 2) by soldiers through their personal stories and their representations; and 3) through visual images of the battlefield or scholarly, journalistic, or fictional accounts of war. Both contributions of this final section present us narratives of how agents and societies – American ones in both cases – try to live through war and how it transforms, refers to, and uses the national body/space. They question our role as citizens, scholars, witnesses, and 'observers' of war.

WHO'S BODIES?

In the first account, looking at the military from the viewpoint of a sociologist of war and the military, Florian Olsen calls on International Relations scholars and geographers to follow the path of historians in studying societies at war and the individuals produced by war. Through a Bourdieusian approach to national habitus, he recounts the American soldier's experience of the war as part of a crisis in an 'imperial society'. In openly discussing the economy of sacrifice of the US national military force from its sociological viewpoint, he dissects the social body politic of the state, thus making sure that the social wounds that are at the very heart of the US national historical memories of war are remembered, known, and that the bodies of those who fight for the US nation are not lost in the dominant national identity narratives.[63] Olsen hence highlights how historical memories of wars are political acts and battlefields in themselves, and argues that a specific geography of sacrifice has resulted from the war in Iraq effort in the US.

In his article, he examines the sociological basis of those fighting wars for the US 'empire', taking the US national armed forces as his case study. Using a comparative research programme developed by French social historian Christophe Charle on the study of European 'imperial societies' of the twentieth century (France, Germany, and Great Britain), he looks at the US through the lens of the concept of an 'imperial society.' As he explains, the crises that an imperial society undergoes translate into specific social wounds (race, class, and gender struggles), cultural attitudes (common stereotypes about the outside world, collective experiences of social and moral anxiety, fears of decline) and painful political struggles over historical memory. Drawing a link between the Vietnam War and the Iraqi War through a social history framework that takes into account the social and racial composition of the US national armed forces and the link to the economy of sacrifice enacted by the military service, Olsen highlights how these wounds resurface at times and how they are linked to an understanding of war as a central feature of US national identity, a sign of America's ever-increasing militarisation of everyday life. In an attempt to bridge together the gulf between competing narratives of US national history, Olsen aims to recover or retrieve what is lost in looking at the experience of war *on* the battlefield without looking at what happens beyond it – and what (national habitus, context, and/or structure) allowed such a caveat in the first place.

NATIONAL BODIES LIVING THROUGH WAR

An excellent rejoinder to Olsen, Tina Managhan explores the domestic ramifications inherited from the geopolitics of war abroad by bringing to the surface the palpable tensions of war beyond the battlefield that made possible what could be conceived of as a geography of support/dissent for/from the war effort. In effect, she explores the psychic space of the subject involved in the feeling of dissent, showing how citizens are interpellated as citizen-soldiers, that is, as citizens who feel it is their duty to morally oppose the war *while* at the same time not preventing the troops from doing their job and staying safe. There is in this sense a militarisation of one's inner space, as the social space of a nation at war clashes with one's psychic space.[64] She exposes how a war of words took place on the home front/national space, producing a geography of intimacy/distance that left those protesting the war to become un*willing* supporters of the war effort through its national bodies to be sacrificed (the 'national soldiers').[65]

On many levels, her exposé is in line with Deborah Cowen's work on militarism and military workfare, where she insists on a geographical explanation of 'military service as a form of labour and citizenship [that] allows us to trace the struggles at home that support or subvert wars abroad'.[66] National soldiers' bodies are (to be) sacrificed outside the home space on the battlefield and then again *beyond* the battlefield, on the home/national space, as the citizenship terrain becomes the national space of war where soldiers' 'excluded inclusion' is written out from the national body. As Cowen points out, 'soldiers are expelled from citizenship in order not to disrupt the practice and meaning of democratic politics, [while] . . . the soldier is also an essential figure to the national state, whose historical and current configuration is premised on the soldier's excluded inclusion'.[67] In a similar vein, Managhan uncovers an 'emotional geography' of antiwar protest that anchors both 'the spatiality and temporality of emotions' to places and bodies (notably those that can be shown to boost the war mobilisation and those that are *made absent* and which she tries to render visible beyond the battlefield); therefrom, she may expose the radical im/possibility of dissent in the United States in regard to the War in Iraq and the call to support troops under George W. Bush.[68] To achieve this, she powerfully puts forth how the discursive conditions of dissent are affected by the figure of the grieving mom, personified by Cindy Sheehan, which brings a simultaneous psychic identification and disavowal of loss that align citizens with the antiwar protester and the nation to support the troops, even in dissent, thus crystallising the metaphor of the 'familial imaginary'[69] enacted by the 'paternalistic' trope of the national security state.[70]

Through losses that one cannot really speak about, the narratives one will use to express one's dissent will thus be limited, as citizens who wish to dissent from the war, especially in the face of its apparent failure, feel compelled to show a certain restraint. In these 'dangerous times', the 'in-between space' of dissent is contested and constrained, and it takes some courage to experience it:

> The dissenter is neither conformist nor revolutionary. She is at once within, but outside of, the community and its conventions. In part because of her liminality the dissenter is often accused of disloyalty and subject to sanction and stigma by state and society. . . . When the physical security of the community of which the dissenter is a member seems jeopardized, these tendencies and temptations intensify. At such

times the critic, the naysayer, the resister, who ordinarily is not welcomed warmly, comes under intense pressure to evacuate the space of dissent, to take sides, to choose allegiance over authenticity.[71]

While Olsen is able to explain how supporting the war has been part of the nationalist logic that was central to the political field (especially for elites around the George W. Bush administration), Managhan exposes how difficult it was for *all* Americans, even those who were political adversaries of the Bush administration, because the discursive rules were well laid out in the political field, as were the costs of contesting them. It seems rather difficult for any antiwar protester to garner support if one does not legitimise one's dissent by expressing – and one must *really say it* – one's support for the troops; in fact, it appears impossible to go beyond this script.

Managhan is indeed exploring the inhabitable and uninhabitable spaces of societies at war in criticising this compelled utterance of support for the troops. It brings to mind the common saying that the Vietnam War was lost at home, with the war becoming unpopular for the American public. She therefore evokes the 'ghosts of Vietnam' to refer to the Vietnam Veterans opposing the war and she exposes well how political and disruptive their (in)visibility in the public space surrounding the war debate is contrary to the Gulf War veterans, who were seen as having killed these 'ghosts' (the Gulf War was said to have overcome the so-called 'Vietnam syndrome' that had affected subsequent military interventions). For conservatives and especially neoconservatives, after September 11, 2001, it was rather important that spaces for dissent be as small as possible, if not muted entirely, in order to prevent 'another Vietnam'. With another disavowed loss in mind, the September 11, 2001 victims, these hawkish figures set the tone for the expression of a masculinist patriotism and a militarised citizenry that establish the conditions of *im*possibility of dissent, laying bare its dangers. Obviously, the Iraqi War, with its high death toll of US troops, only made this harder.

So it is through a public figure such as Cindy Sheehan, the dead soldier's mother turned into an antiwar activist as the nation's emblematic grieving mother, that the antiwar movement was able to inscribe its dissent in a narrative that would be more acceptable to mainstream America. If it can be levelled that 'there was a healthy appetite for both 'high' and 'low' cultural forms that criticised the Bush administration's foreign policy', it would be faulty to assert that space for dissent was easily created.[72] It is with that difficult societal context in mind that Managhan explains how Sheehan became the public legitimate voice of dissent, her body being felt psychologically as the nation's through a process of 'double recognition' that establishes one subject as an iconic figure. She thus explores Judith Butler's 'psychic life of power' in affirming dissent and support, as social subjects are formed through their emotions: if Sheehan's loss was personal, the nation's was political. Coincidentally, this is what allowed a space for dissent – through the support for the troops and their families – and, in return, this is how war dissenters were self-disciplined as citizens and as the nation.

On the Special Issue: Struggling to Come to Terms with the War on Terror in the 'Land of Plenty'

In hindsight, one could say that this special issue project follows Wim Wenders's 2004 movie, *Land of Plenty*, as it aims to represent the War on Terror as lived experience and to offer a space for critical reflection on the war enterprise *beyond* the battlefield. In many ways, the conclusion of the movie echoes the difficulties we all have – but especially the Obama administration – in coming to terms with the War on Terror as the most recent past of America. Hence, in *Land of Plenty*, two main characters, Paul and Lana, an uncle and a niece, played respectively by John Diehl and Michelle Williams, are seen trying to cope, each in their own way, with the September 11, 2001 tragedy. Lana, who is twenty years old, is just coming back to her country, after having lived abroad with her missionary parents for most of her life. On the one hand, the movie puts on the misguided patriotism and the paranoia of a Vietnam War veteran (Paul) who wishes to make amends for his country's failure that led to the fall of the twin towers by trying to debunk the next 'sneaky' attack. On the other hand, through Lana's experience of living in the West Bank and in Africa and her experience of the September attacks, the movie ends up as a powerful antiwar intervention that wishes to make those lives lost on the morning of September 11, 2001 mean something and be heard.

Inasmuch as this special issue got us to address the other spaces of war, those beyond the battlefield, the ending scene of this movie plunges us in the dis/comforting geography of dissent/support to which we alluded in discussing Tina Managhan's emotional geography contribution to the War on Terror. Our loyalties are necessarily displaced, our securities disaffected, and our convictions disrupted by our collective exploration of war *beyond* the battlefield. As one of the closing moments of the films evokes, in a dialogue between the two relatives as they are reflecting on the meaning of the fall of the two towers for them, we have an obligation to make a space for those whose voices are being shut out by the drums of war and by the dominant narrative self-imposed by the rhetoric of the War on Terror.

> [Lana]: I was so far away that it was nighttime . . . . . . when the towers fell. And at first, I didn't . . . I didn't know what was going on. I just heard people . . . cheering.
>
> [Paul]: Cheering?

[Lana]: There was a crowd gathered out in the street close to where we lived.

[Paul]: Terrorists.

[Lana]: No. No. I think that's the part that hurts the most. They weren't – they were just . . . . . . just ordinary people.

[Paul]: Why in the hell were they cheering?

[Lana]: Because they hate us. Because they hate America. And it came from such an honest place, and from so many people that I just – I knew that something had gone wrong and . . . and that's – that's become my nightmare.

[Paul]: Over 3,000 civilians were killed on that day. They were innocent people.

[Lana]: Yeah, and – and it's their voices that I need to hear. Because I really don't think that they would want any more people killed in their name.

(Moments later, in front of the Ground Zero site in New York city, Lana, still referring to the victims of September 11, 2001, says to her uncle: 'Let's be quiet. Let's just try and listen . . .'.)

(excerpt from Wim Wenders' *Land of Plenty*)

Following the difficulties raised in the process of *Vergangenheitbewältigung* (coming to terms with the past) and which the Obama administration now faces in seeking to let go of the Bush legacy of the War on Terror, this special issue wishes to re-historicise the War on Terror beyond the battlefield, to tell (other) stories of war, and write new histories of the War on Terror – by voicing silences and seeing those who have experienced war without seeing the battlefield. The anticipated added value and originality of this collection of essays therefore resides in its commitment to disrupt both a complacent or outright critique of the War on Terror as it unfolded in Iraq, in America, and elsewhere by (re)presenting war beyond the battlefield in addressing issues still related to but not limited to the actual waging of war. The great themes of this special issue are meant as a guideline for readers and as a general glance at the sociological practices and the politics of security that considering the transversalities of the experience of war beyond the battlefield the War on Terror generates. The intent is to fill a void that a geopolitical and strategic literature has long neglected and to see the War on Terror transversally as social relations, as politics, as lived real and virtual experiences, and as imagined and represented realities.

ACKNOWLEDGEMENTS

I want to thank the editors of *Geopolitics*, anonymous reviewers, and all contributors to this special issue for making it happen. I personally wish to thank Simon Dalby for not only steering the whole process and bringing it to its completion, but for firmly believing in it in the first place. Special mentions go to Anne-Marie D'Aoust for reviewing the introduction piece.

NOTES

1. President B. Obama, quoted in M. Hirsh, 'Memo to President Obama. Never Mind Iraq. Just End the War on Terror', *Newsweek*, 21 Feb. 2008, available at <http://www.newsweek.com/id/114385>, accessed 10 March 2008.

2. P. Virilio, quoted in an interview with J. Der Derian (ed.), *The Virilio Reader* (Malden, MA: Blackwell 1998) p. 17 (our emphasis).

3. S. H. Jones and D. B. Clarke, 'Waging Terror: The Geopolitics of the Real', *Political Geography* 25 (2006) pp. 298–314.

4. R. Pain and S. Smith (eds.), *Fear: Critical Geopolitics and Everyday Life* (Aldershot: Ashgate 2008).

5. K. Dodds and A. Ingram (eds.), *Spaces of Security and Insecurity: Geographies of the War on Terror* (Aldershot: Ashgate 2009).

6. R. Pain and S. Smith, 'Preface', in Pain and Smith (note 4) p. xv.

7. K. Dodds and A. Ingram, 'Spaces of Security and Insecurity: Geographies of the War on Terror', in Dodds and Ingram (note 5) p. 10.

8. Worth mentioning here are: R. Bleiker, *Aesthetics and World Politics* (New York: Palgrave Macmillan 2009); S. Croft, *Culture, Crisis and America's War on Terror* (Cambridge, Cambridge University Press 2006); F. Debrix and M. Lacy (eds.), *Geopolitics of American Insecurity: Terror, Power, and Foreign Policy* (London: Routledge 2009); F. Debrix, *Tabloid Terror: War, Culture, and Geopolitics* (New York: Routledge 2008); J. Der Derian, *Virtuous War: Mapping the Military-Industrial-Media-Entertainment Network*, 2nd ed. (London and New York: Routledge 2009); K. Dodds, 'Steve Bell's Eye: Cartoons, Geopolitics and the Visualization of the "War on Terror"', *Security Dialogue* 38/2 (2007) pp. 157–177; K. Dodds, ' "Have You Seen Any Good Films Lately?" Geopolitics, International Relations and Film', *Geography Compass* 2/2 (2008) pp. 476–494; K. Dodds, 'Screening Terror: Hollywood, the United States and the Construction of Danger', *Critical Studies on Terrorism*, 1/2 (2008) pp. 227–243; M. Doucet, 'Child's Play: The Political Imaginary of International Relations and Contemporary Popular Children's Films', *Global Society* 19/3 (2005) pp. 289–306; K. Grayson, M. Davies and S. Philpott, 'Pop goes IR? Researching the Popular Culture–World Politics Continuum', *Politics* 29/3 (2009) pp. 155–163; R. Hughes, 'Through the Looking Blast: Geopolitics and Visual Culture', *Geography Compass* 1/5 (2007) pp. 976–994; R. Jackson, *Writing the War on Terrorism: Language, Politics and Counter-Terrorism* (Manchester: Manchester University Press 2005); T. Juneau and M. Sucharov, 'Narratives in Pencil: Using Graphic Novels to Teach Israeli-Palestinian Relations', *International Studies Perspectives* 11/2 (2010) pp. 172–183; A. Kangas, 'From Interfaces to Interpretants: A Pragmatist Exploration into Popular Culture as International Relations', *Millenium* 38/2 (2009) pp. 317–343; D. Kellner, *Cinema Wars: Hollywood Film and Politics in the Bush-Cheney Era* (Malden, MA: Wiley-Blackwell 2010); B. J. Muller, 'Securing the Political Imagination: Popular Culture, the Security *Dispositif* and the Biometric State', *Security Dialogue* 39/2–3 (2008) pp. 199–220; D. H. Nexon and I. B. Neumann (eds.), *Harry Potter and International Relations* (Lanham,

MD: Rowman & Littlefield 2006); M. Power and A. Crampton, 2005, 'Reel Geopolitics: Cinemato-graphing Political Space', *Geopolitics* 10/2 (2005) pp. 193–203; M. Power, 'Digitized Virtuosity: Video War Games and Post-9/11 Cyber-Deterrence', *Security Dialogue* 38/2 (2007) pp. 271–288; M. J. Shapiro, *Cinematic Geopolitics* (New York: Routledge 2009); C. Weber, *Imagining America at War: Morality, Politics, and Film* (London and New York: Routledge 2006).

9. J. Dittmer, *Popular Culture, Geopolitics, and Identity* (Lanham, MD: Rowman & Littlefield Publishers, Inc. 2010).

10. J. Dittmer, quoted in 'Popular Geopolitics, Culture and Representations – Jason Dittmer: Interview with Leonhardt van Efferink', June 2010, available at <http://www.exploringgeopolitics.org/ Interview_Dittmer_Jason_Popular_Geopolitics_Culture_Identity_Representations_Affect_Postcolonialism _Social_Constructivism_New_Media_Americanisation_James_Bond.html>.

11. M. Dillon and J. Reid, *The Liberal Way of War: Killing to Make Life Live* (London and New York: Routledge 2009).

12. F. Macdonald, R. Hughes, and K. Dodds, 'Envisioning Geopolitics', in F. Macdonald, R. Hughes, and K. Dodds (eds.), *Observant States: Geopolitics and Visual Culture* (London and New York: I.B. Tauris 2010) p. 5.

13. S. Dalby, 'Critical Geopolitics and Security', in P. Burgess (ed.), *Handbook of New Security Studies* (London: Routledge 2009) p. 51.

14. B. Klein, quoted in J. George, *Discourses of Global Politics: A Critical (Re)Introduction to International Relations* (Boulder, CO: Lynne Rienner 1994) p. 30.

15. Dittmer (note 9) p. 94.

16. F. Wilmer, '"Ce n'est pas une guerre/This Is Not a War": The International Language and Practice of Political Violence', in F. Debrix (ed.), *Language, Agency, and Politics in a Constructed World* (Armonk, NY, and London: M.E. Sharpe 2003) p. 222.

17. Ibid., p. 221. See also R. Jackson, *Writing the War on Terrorism: Language, Politics and Counter-Terrorism* (Manchester: Manchester University Press 2005).

18. J. Der Derian, 'Imaging Terror: Logos, Pathos and Ethos', *Third World Quarterly* 26/1 (2005) p. 27.

19. G. Cheeseman, 'Military Force(s) and In/security', in K. Booth (ed.), *Critical Security Studies and World Politics* (Boulder, CO, and London: Lynne Rienner 2005) p. 82.

20. M. V. Rasmussen, '"It Sounds Like a Riddle": Security Studies, the War on Terror and Risk', *Millennium: Journal of International Studies* 33/2 (2004) p. 382 (our emphasis).

21. L. Amoore and M. de Goede (eds.), *Risk and the War on Terror* (New York and London: Routledge 2008); M. de Goede, 'Beyond Risk: Premediation and the Post-9/11 Security Imaginary', *Security Dialogue* 39/2–3 (2008) pp. 155–176.

22. M. Foucault, 'La gouvernementalité' [The governmentality], in *Dits et écrits II, 1976–1988* (Paris: Quarto Gallimard 2001 [orig. 1978]) p. 655.

23. Whereas the idea of sovereignty, derived from the medieval sovereign, whose authority was absolute and omnipotent, had provided the initial framework for the emergence of the modern state, the logic of 'governmentality' came along to deal with the historical evolution of the 'population' during the eighteenth century, a direct consequence of the simultaneous growth of industrialisation, of production, and, consequently, of consumption, as well as the demographic expansion and urbanisation. The population emerged 'both as an object of study, circumscribed by new forms of knowledge (such as statistics), and as the 'thing' in need of being governed – and thus as the point of convergence of knowledge and power.' C. Epstein, 'Guilty Bodies, Productive Bodies, Destructive Bodies: Crossing the Biometric Borders', *International Political Sociology* 1/2 (June 2007) p. 151.

24. J. Coaffee, *Terrorism, Risk, and the Global City: Towards Urban Resilience* (Aldershot: Ashgate 2009).

25. Hence, as homeland security lies not only, or not *merely*, within the purview of the state, as the private sector assumes a big share of the 'homeland security market', homeland security consumes power from the state as it is taking shape through discourses, procedures, institutions, and practices so as to become a new 'homeland security-industrial complex'.

26. C. Aradau and R. Van Munster, 'Exceptionalism and the "War on Terror": Criminology Meets International Relations', *British Journal of Criminology* 49 (2009) pp. 686–701; J. McCulloch and S. Pickering, 'Pre-Crime and Counter-Terrorism: Imagining Future Crime in the "War on Terror"', *British Journal of Criminology* 49 (2009) pp. 628–645 ; G. Mythen and S. Walklate, 'Criminology and Terrorism: Which Thesis? Risk Society or Governmentality?', *British Journal of Criminology* 46 (2006) pp. 379–398.

27. C. Aradau and R. Van Munster, 'Governing Terrorism Through Risk: Taking Precautions, (Un)Knowing the Future', *European Journal of International Relations* 13/1 (Jan. 2007) pp. 90–91.

28. One similar conceptualisation of this idea of the War on terror *beyond* the battlefield, when societies and individuals are turned into (un)wilful participants of this war is the Deleuzian and Guattarian concept of the 'war machine'. F. Debrix, *Tabloid Terror: War, Culture, and Geopolitics* (London and New York: Routledge 2008) p. 120. See also the special issue on Deleuze and War in *Theory & Event* in 2010.

29. R. Stangel, quoted in M. Andrejevic, *iSpy: Surveillance and Power in the Interactive Era* (Lawrence, KS: University Press of Kansas 2007) pp. 4–5, 163.

30. R. Santorum, quoted in Andrejevic (note 29) p. 164.

31. Andrejevic (note 29) pp. 164–165.

32. C. H. Gray, *Peace, War, and Computers* (New York and London: Routledge 2005) p. 40.

33. Ibid.

34. R. Bleiker, *Popular Dissent, Human Agency, and Global Politics* (Cambridge: Cambridge University Press 2000) p. 2.

35. D. Campbell, 'Political Prosaics, Transversal Politics', in M. J. Shapiro and H. Alker (eds.), *Challenging Boundaries: Global Flows, Territorial Identities* (Minneapolis: University of Minnesota Press 1996) p. 24.

36. Ibid., p. 23 (original emphasis; in bold characters, our emphasis).

37. C. Lutz, *Homefront: A Military City and the American Twentieth Century* (Boston: Beacon Press 2002) pp. 2, 8.

38. Haunting, as Derrida uses it in *The Specters of Marx*, refers to a nagging, constant obsession, fear, and fixed idea (it is the translation of Derrida's term *hantise*); P. Kamuf, in J. Derrida, *Specters of Marx: The State of the Debt, the Work of Mourning, & the New International*, trans. by P. Kamuf (New York and London: Routledge 1994) p. 177, n. 2.

39. M. Ferro, *La Grande Guerre 1914–1918* [*The Great War 1914–1918*] (Paris: Gallimard 1990) p. 55.

40. Ibid., p. 398 (the author's translation).

41. M. de Larringa and M. Doucet, 'Introduction: The Global Governmentalization of Security and the Securitization of Global Governance', in M. de Larringa and M. Doucet (eds.), *Security and Global Governmentality: Governance, Globalization and the State* (London: Routledge 2010).

42. Ferro (note 39) p. 55.

43. H. L. Dreyfus and P. Rabinow, 'What Is Maturity? Habermas and Foucault on "What Is Enlightenment?"', in D. Couzens Hoy (ed.), *Foucault: A Critical Reader* (Oxford and Cambridge: Blackwell 1986) pp. 115–117.

44. Discussing port security and issues relating the space of logistics, Deborah Cowen will indeed talk about the rise of 'homeland cities'. D. Cowen, 'Containing Insecurity: US Port Cities and the "War on Terror"', in S. Graham (eds.), *Disrupted Cities: When Infrastructure Fails* (New York and London: Routledge 2009) p. 71.

45. US Department of Defense, CENTCOM, 2007; see the Area of responsibility map, available at <http://www.centcom.mil/en/map-mideast/index.php>, accessed 10 Nov. 2007.

46. J. Coaffee and D. M. Wood, 'Security is Coming Home: Rethinking Scale and Constructing Resilience in the Global Urban Response to Terrorist Risk', *International Relations* 20/4 (2006) p. 514.

47. J. Morrissey, 'Liberal Lawfare and Biopolitics: US Juridical Warfare in the War on Terror', this Special Issue.

48. Ibid.

49. G. Agamben, *State of Exception*, trans. by K. Attell (Chicago: University of Chicago Press 2005).

50. D. Gregory, 'Vanishing Points: Law, Violence and Exception in the Global War Prison', in D. Gregory and A. Pred (eds.), *Violent Geographies: Fear, Terror and Political Violence* (New York: Routledge 2007) pp. 205–236.

51. Morrissey (note 47).

52. D. Grondin, 'The New Frontiers of the National Security State: The US Global Governmentality of Contingency', in M. Doucet and M. de Larringa (eds.), *Security and Global Governmentality: Globalization, Governance and the State* (London: Routledge 'PRIO New Security Studies', 2010) pp. 79–95.

53. J. Masco, *The Nuclear Borderlands: The Manhattan Project in Post-Cold War New Mexico* (Princeton and Oxford: Princeton University Press 2006).

54. J. Hunner, *Inventing Los Alamos: The Growth of an Atomic Community* (Norman, OK: University of Oklahoma Press 2001) pp. 11, 236.

55. N. Curtis, *War and Social Theory: World, Value and Identity* (New York: Palgrave 2006) p. 134.

56. D. Grondin and P. Racine-Sibulka, ' "It's Not Science Fiction; It's What We Do Every Day!": The Cinematics of War Videogames, or Selling War as Game', paper presented at the Association of American Geographers, Washington, DC, 14–18 April 2010 [under review at *Review of International Studies*].

57. M. Salter, 'The Geographical Imaginations of Video Games: *Diplomacy, Civilization, America's Army* and *Grand Theft Auto IV*', this Special Issue.

58. S. Graham, *Cities Under Siege: The New Military Urbanism* (London: Verso 2009) p. xv.

59. S. Graham, *Disrupted Cities: When Infrastructure Fails* (London and New York: Routledge 2010) p. 24.

60. S. Graham, 'Combat Zones that See: Urban Warfare and US Military Technology', in F. Macdonald, R. Hughes, and K. Dodds (eds.), *Observant States: Geopolitics and Visual Culture* (London and New York: I.B. Tauris 2010) p. 206; Dillon and Reid (note 11) pp. 121–122.

61. Gray (note 32).

62. Salter (note 57).

63. W. Hixson, *The Myth of American Diplomacy: National Identity and US Foreign Policy* (New Haven: Yale University Press 2008).

64. J. Orr, 'The Militarization of Inner Space', *Critical Sociology* 30/2 (2004) pp. 451–481.

65. D. Cowen, 'From the American Lebensraum to the American Living Room: Class, Sexuality, and the Scaled Production of "Domestic" Intimacy', *Environment and Planning D: Society and Space* 22 (2004) pp. 755–771.

66. D. Cowen, 'National Soldiers and the War on the Cities', *Theory & Event* 10/2 (2007).

67. Ibid.

68. L. Bondi, J. Davidson, and M. Smith, 'Introduction: Geography's "Emotional Turn"', in J. Davidson, L. Bondi, and M. Smith (eds.), *Emotional Geographies* (Aldershot: Ashgate 2007) p. 3.

69. This term is borrowed from D. Cowen and E. Gilbert, 'Citizenship in the "Homeland": Families at War', in D. Cowen and E. Gilbert (eds.), *War, Citizenship, Territory* (New York: Routledge 2008) p. 262.

70. I. M. Young, 'The Logic of Masculinist Protection: Reflections on the Current Security State', *Signs* 29 (2003) pp. 1–26.

71. A. Sarat, 'Terrorism, Dissent, & Repression: An Introduction', in A. Sarat (ed.), *Dissent in Dangerous Times* (Ann Arbor, MI: University of Michigan Press 2005) p. 2

72. If the late leftist American historian Howard Zinn rightly pointed out that in times of war, 'when societies are mesmerized by a culture of fear and soothed by the collective buzz of patriotism' that the most patriotic act ordinary citizens could do was to criticise their government, he certainly erred in believing that it was artists who were in the best position to resist the government. Debbie Lisle persuasively argued how narrow the 'Zinnian' view of dissenting art as higher patriotism was, as artists faced the same contextual dilemma that other 'ordinary' citizens were confronted with: 'How could they be *critical* of the USA when it had suffered such a devastating loss? How could the tradition of 'dissenting' art say anything meaningful in such conditions of grief?' D. Lisle, 'Benevolent Patriotism: Art, Dissent and *The American Effect*', *Security Dialogue* 38/2 (June 2007) pp. 234, 238.

Liberal Lawfare and Biopolitics: US Juridical Warfare in the War on Terror

JOHN MORRISSEY

Department of Geography, National University of Ireland, Galway, Ireland

Two basic forms of 'lawfare' are employed by the United States in its enactment of the war on terror, both of which have a biopolitical focus. The first strategy has been well documented.[1] It involves the indefinite detention and sometimes extraordinary rendition of enemy combatants, legally sanctioned and politically justified by the 'exceptional' circumstances of late modern war and terrorist violence. Geography plays a central role in strategy number one: the legal statuses of detainees, whose lives and bodies are cast out and denied basic juridical rights, are bounded, identified and allowed for in extra-territorial spaces throughout the world, from Guantanamo Bay to Bagram Air Force Base. Such exceptional biopolitical spaces are essentially 'defensive' and operate at the local scale. On the contrary, the second seldom-discussed legal strategy conditions and protects the US military in 'offensive' mode, operates at the national and transnational scale, and involves the careful legal designation and protection of US military personnel in forward deployed areas.[2] This paper is centrally concerned with strategy number two – a strategy that can be defined as 'forward juridical warfare' and involves the US military's mobilisation of the law in the waging of war along the 'new frontiers' of its war on terror. The paper seeks to expound the legal and biopolitical constitution and operation of the current US military's forward presence overseas, and begins by drawing on recent work on biopolitics that has sought in various ways to critique the proliferation of practices of liberal lawfare and securitization in our contemporary world.

At the Direction of the President, we will defeat adversaries at the time, place, and in the manner of our choosing – setting the conditions for future security.[3]

— *National Defense Strategy of the United States of America*, 2005

For too long, the law has not been understood as a critical instrument of foreign policy.[4]

— Harvey Rishikof, former Chair, Department of National Security Strategy, National War College, 2008

If, therefore, conclusions can be drawn from military violence . . . there is inherent in all such violence a lawmaking character.[5]

— Walter Benjamin, *Critique of Violence*, 1921

Security is a way of making the old armatures of the law and discipline function.[6]

— Michel Foucault, Lecture at the Collège de France, 1978

FOUCAULT'S 'SOCIETY OF SECURITY': TOWARDS A BIOPOLITICAL CRITIQUE OF THE WAR ON TERROR

What does it mean to place 'life' at the centre of political inquiry? What is achieved in this move? Is it possible to speak of a 'spatiality of biopolitics', which not only pays attention to scale but also to the complex constellation of sovereign and biopolitical power? And if so, is it helpful any longer to speak of 'biopolitics' and 'geopolitics' separately? In recent years, such questions have featured prominently in reflections across the social sciences that have sought to theorise the relations between 'territory' and 'sovereign power', 'populations' and 'biopolitical power'.[7] Though 'biopolitics' is a much contested term, deployed differently in an array of contexts, a key focus of examination in the increasingly extensive literature on the subject has nonetheless been on the spatial politics through which life is constituted and governed; in other words, how life is incorporated into modern forms of governmentality. And this, of course, is a particularly geographical concern.

For geographers, putting 'life' at the centre of political critique poses a number of intriguing theoretical challenges that have been taken up in a variety of ways.[8] Many have drawn on the work of Foucault and particularly his recently translated lectures on security, territory and population at the Collège de France in 1978.[9] For Foucault, geography was, of course, pivotal to his thinking on biopower;[10] and, together with the figure of 'population', was central too in the advancement of what Derek Gregory has called his "biopolitical imaginary".[11] In *Security, Territory, Population*, Foucault outlines what he saw as the eighteenth-century[12] move from power primarily directed over 'territory' to power increasingly focused on 'population'. From this point, Foucault argues a new "general economy of power" began to emerge, dominated by "mechanisms" or "technologies" of security whose endgame was the identification, regulation and circulation of populations; and this new "society of security" was enabled by "making the old armatures of the law and discipline function".[13] The new regulatory "apparatuses (*dispositifs*) of security" reflected for Foucault a shift in the sovereign's concerns from: "the safety of his territory" to the "security of the population"; from "what limit to impose" to "facilitating the proper circulation of people"; from traditional "sovereign power" to modern "biopolitical power".[14] Foucault's outlining of the governmental shift towards the security and securitization of whole "populations" is especially instructive to the argument I want to make later concerning the biopolitical strategies of US military commanders on the new frontiers of the war on terror. For commanders, the 'population' under their command – including especially US troops – presents a dialectic of what Foucault calls "juridical-political" subjects and "technical-political" objects of "management and government".[15]

In theorising the confluence of 'security, territory, population' in early 1978, Foucault introduced the concept of 'governmentality' for the first time, and indeed acknowledged his preference for "a history of "governmentality"" as a more apposite title for his lecture course that spring. 'Governmentality' was formulated as both a "problematic" that marks the entry of the "modern state in a general technology of power", but also as an analytical tool that involves a "methodological principle" of going behind or outside the 'state' (a move away, in other words, from an "institutional-centric" approach) to conceive of a wider perspective on "the technology of power".[16] 'Governmentality', for Foucault, is understood first and foremost as an assemblage of "institutions, procedures, analyses and reflections, calculations, and tactics" that capacitate a form of power that "has the population as its target, political economy as its major form of knowledge, and apparatuses of security as its essential technical instrument".[17] And the era of modernity is marked not by the state's "takeover" of society but rather by how the state became gradually "administrative" and "governmentalized" and "controlled by apparatuses of security".[18]

Foucault's envisioning of a more governmentalised and securitized modernity, framed by a ubiquitous architecture of security, speaks on various levels to the contemporary US military's efforts in the war on terror, but I want to mention three specifically, which I draw upon through the course of the paper. First, in the long war in the Middle East and Central Asia, the US military actively seeks to legally facilitate both the 'circulation' and 'conduct' of a target population: its own troops. This may not be commonly recognized in biopolitical critiques of the war on terror but, as will be seen later, the Judge Advocate General Corps has long been proactive in a 'juridical' form of warfare, or lawfare, that sees US troops as 'technical-biopolitical' objects of management whose 'operational capabilities' on the ground must be legally enabled. Second, as I have explored elsewhere, the US military's 'grand strategy of security' in the war on terror – which includes a broad spectrum of tactics and technologies of security, including juridical techniques – has been relentlessly justified by a power/knowledge assemblage in Washington that has successfully scripted a neoliberal political economy argument for its global forward presence.[19] Securitizing economic volatility and threat and regulating a neoliberal world order for the good of the global economy are powerful discursive touchstones registered perennially on multiple forums in Washington – from the Pentagon to the war colleges, from IR and Strategic Studies policy institutes to the House and Senate Armed Services Committees – and the endgame is the legitimisation of the military's geopolitical and biopolitical technologies of power overseas.[20] Finally, Foucault's conceptualisation of a 'society of security' is marked by an urge to 'govern by contingency', to 'anticipate the aleatory', to 'allow for the evental'.[21] It is a 'security society' in which the very language of security is promissory, therapeutic and appealing to liberal improvement. The lawfare of the contemporary US military is precisely orientated to plan for the 'evental', to anticipate a series of future events in its various 'security zones' – what the Pentagon terms 'Areas of Responsibility' or 'AORs' (see Figure 1).[22] These AORs equate, in effect, to what Foucault calls "spaces of security", comprising "a series of possible events" that must be securitized by inserting both "the temporal" and "the uncertain".[23] And it is through preemptive juridical securitization 'beyond the battlefield' that the US military anticipates

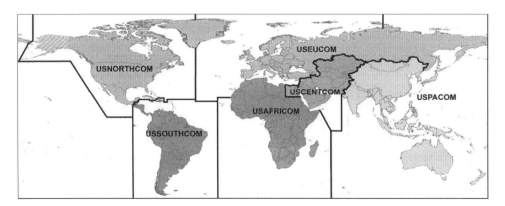

FIGURE 1 "The World with Commanders' Areas of Responsibility", 2010[24] (color figure available online).

and enables the necessary biopolitical modalities of power and management on the ground for any future interventionary action.

AORs AND THE 'MILIEU' OF SECURITY

For CENTCOM Commander General David Petraeus, and the other five US regional commanders across the globe, the 'population' of primary concern in their respective AORs is the US military personnel deployed therein. For Petraeus and his fellow commanders, US ground troops present perhaps less a collection of "juridical-political" subjects and more what Foucault calls "technical-political" objects of "management and government".[25] In effect, they are tasked with governing "spaces of security" in which "a series of uncertain elements" can unfold in what Foucault terms the "milieu".[26] What is at stake in the 'milieu' is "the problem of circulation and causality", which must be anticipated and planned for in terms of "a series of possible events" that need to "be regulated within a multivalent and transformable framework".[27] And the "technical problem" posed by the eighteenth-century town planners Foucault has in mind is precisely the same technical problem of space, population and regulation that US military strategists and Judge Advocate General Corps (JAG) personnel have in the twenty-first century.

For US military JAGs, their endeavours to legally securitize the AORs of their regional commanders are ultimately orientated to "fabricate, organize, and plan a milieu" even before ground troops are deployed (as in the case of the first action in the war on terror, which I return to later: the negotiation by CENTCOM JAGs of a Status of Forces Agreement with Uzbekistan in early October 2001).[28] JAGs play a key role in legally conditioning the battlefield, in regulating the circulation of troops, in optimising their operational capacities, and in sanctioning the privilege to kill. The JAG's milieu is a "field of intervention", in other words, in which they are seeking to "affect, precisely, a population".[29] To this end, securing the aleatory or the uncertain is key. As Michael Dillon argues, central to the securing of populations are the "sciences of the aleatory or the contingent" in which the "government of population" is achieved by the sciences of "statistics and probability".[30] As he points out elsewhere, you "cannot secure anything unless you know what it is", and therefore securitization demands that "people, territory, and things are transformed into epistemic objects".[31] And in planning the milieu of US ground forces overseas, JAGs translate regional AORs into legally enabled grids upon which US military operations take place. This is part of the production of what Matt Hannah terms "mappable landscapes of expectation";[32] and to this end, the aleatory is anticipated by planning for the 'evental' in the promissory language of securitization.

The ontology of the 'event' has recently garnered wide academic engagement. Randy Martin, for example, has underlined the evental discursive underpinnings of US military strategy in the war on terror; high-lighting how the risk of future events results in 'preemption' being the tactic of their securitization.[33] Naomi Klein has laid bare the powerful event-based logic of 'disaster capitalism';[34] while others have pointed out how an ascendant 'logic of premediation', in which the future is already anticipated and "mediated", is a marked feature of the "post-9/11 cultural landscape".[35] But it was Foucault who first cited the import of the 'evental' in the realm of biopolitics. He points to the "anti-scarcity system" of seventeenth-century Europe as an early exemplar of a new 'evental' biopolitics in which "an event that could take place" is prevented before it "becomes a reality".[36] To this end, the figure of 'population' becomes both an 'object', "on which and towards which mechanisms are directed in order to have a particular effect on it", but also a 'subject', "called upon to conduct itself in such and such a fashion".[37] Echoing Foucault, David Nally usefully argues that the emergence of the "era of bio-power" was facilitated by "the ability of 'government' to seize, manage and control individual bodies and whole populations".[38] And this is part of Michael Dillon's argument about the "very operational heart of the security dispositif of the biopolitics of security", which seeks to 'strategize', 'secure', 'regulate' and 'manipulate' the "circulation of species life".[39] For the US military, it is exactly the circulation and regulation of life that is central to its tactics of lawfare to juridically secure the necessary legal geographies and biopolitics of its overseas ground presence.

US FORWARD PRESENCE IN THE WAR ON TERROR: THE ENDURING IMPORT OF 'LAND POWER'

In considering the US military's legal tactics to empower its specifically 'biopolitical project of security' overseas, it is important to first sketch out the recent historical geographical evolution of the contemporary 'milieu' of US ground forces abroad – or what some refer to as the American 'leasehold empire'.[40] In doing so, I want to especially underline the evolving import of 'land power' – defined by 'land access', not territorial control – increasingly identified by military strategists in Washington from the early 1980s, which in turn behoved US military JAGs to foreground the legal terrain of any planned for ground presence. They were tasked with forecasting the evental, forestalling the uncertain; preconfiguring, in other words, the biopolitical modalities of US operations on the ground. And this expressly biopolitical project of securitization works in tandem, of course, with a broader geopolitical and geoeconomic project of securitization. It is the biopolitical enabling of land power.

US grand strategy in the war on terror is chiefly built upon a network of bases in the Middle East and Central Asia, the area militarily managed by the aforementioned US CENTCOM. Its AOR, seen in Figure 2, extends

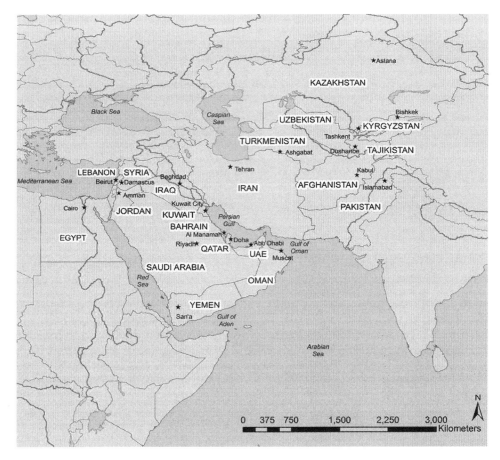

FIGURE 2 CENTCOM Area of Responsibility, 2010 (color figure available online).

from Egypt across the Arabian Gulf to Iraq, Iran and Afghanistan. It calls this vast area the 'Central Region' and, at the beginning of 2010, it had over 225,000 armed services personnel forward deployed therein.[41] Historically, the current extent of both CENTCOM forces and bases represents a high point of US geostrategic presence in the region. However, in the early years of the command's inception in the 1980s, the contemporary scenario could only have been dreamed of by foreign policy strategists in Washington.[42]

In early 1980, apart from the tiny atoll of Diego Garcia in the Indian Ocean, some 2,000 miles southwest of the coast of Oman, the United States held no military bases anywhere in the Middle East, from Turkey to the Philippines. At this juncture, as then Undersecretary of Defense Robert Komer communicated in Congress, Persian Gulf countries "most emphatically do not want formal security arrangements with us".[43] By the time

CENTCOM came into being in 1983, 'prepositioning' at sea of military arsenals, logistical supports and subsistence supplies was the principal US national security strategy overseas.[44] However, a marked concern of contemporary military planners in Washington was that prepositioning on ships was ultimately a limited war strategy on its own; particularly for the new military thinking behind 'rapid deployment forces'. A range of government advisory think tanks in strategic studies lamented the perennial bigger challenge: the 'problem of access'. Jeffrey Record, at the Institute for Foreign Policy Analysis, for example, saw "secure military access ashore in the Persian Gulf" as essential for US foreign policy in the region.[45] For Record, the imperative of 'land access' was clear:

> To get ashore, intervention forces must have access to ports, airfields, and other reception facilities. To stay ashore, they require continued access to proximate logistical support bases.[46]

Such concerns were, of course, driven too by the Cold War geopolitical context of the early 1980s, which saw the prevailing view of "American deterrent capabilities" against the Soviet Union in the Middle East as "frighteningly pessimistic", given that the Soviets had twenty-four infantry divisions based north of Iran, while the US had four divisions "based thousands of miles away" in the continental United States.[47]

Consequently, a proactive Department of Defense strategy for forging military links in the broader Persian Gulf region was vigorously pursued in the early 1980s by first initiating 'joint training exercises' with host countries. Egypt was originally the main target, and the establishment of the military training exercise, *Bright Star*, there in late 1980 was heralded by the Carter administration as the first step in the necessary ground forces access for the US military in the Middle East.[48] The Department of Defense simultaneously pursued a policy of securing 'access rights' for its armed forces with various countries across the Persian Gulf and Horn of Africa through the course of the 1980s (typically via bilateral agreements involving arms sales as a key component);[49] and the growing US military capabilities in the region culminated in the capacity to rapidly respond to Saddam Hussein's invasion of Kuwait in 1990. In the wake of success in the Gulf War, a new US deterrence strategy, which tasked CENTCOM with the military-economic securitization of the Persian Gulf, emerged in the 1990s and this prompted the command to rigorously seek to secure more permanent access rights and extend its basing structure and land prepositioning across the region.[50] By 1994, the US had forged close relationships with all six Gulf Cooperation Countries (GCC), signing broad bilateral defence and access agreements, including classified Status of Forces Agreements, with each state.[51]

The import of 'land access' for the US military's contemporary projects of security overseas continues to be cited in the corridors of power in

Washington today. In their 2005 report to Congress, the US Overseas Basing Commission, for instance, not only reaffirmed the enduring significance of land access for the military, but also divulged the preemptive logic driving current US basing strategy:

> The U.S. overseas basing structure must serve both in the near term and for decades to come . . . any base structuring cannot be designed to deal only with the threats of today. The base structure we develop in the near future must enable us to meet the threats that will emerge over the next quarter century and beyond.[52]

And the commission were at pains to underline, in particular, the imperative of "maintaining a forward presence" for CENTCOM's geopolitical and geoeconomic mission.[53] A year after the review, the then CENTCOM Director of Logistics Major General Brian Geehan outlined how the command supported a staggering 128 operating bases across its AOR;[54] and the concentration of these mirror some of the key nodes in the political economy of oil production in the Persian Gulf. Apart from the operational infrastructure necessary to sustain the ongoing wars in Iraq and Afghanistan, the bulk of CENTCOM's long-term facilities architecture on the ground is in the energy-rich GCC countries of the Arabian Peninsula (see Table 1).

TABLE 1 US access facilities in GCC countries in the Persian Gulf, 2010

Country	Facility[55]	Description
Bahrain	- *Al Manamah Port* (CENTCOM Navy and Marine HQs) - *Mina al-Sulman Port* - *Muharraq Airfield* - *Sheik Isa Air Base*	Bahrain hosts the US Navy's HQ in the Persian Gulf at its capital, Al Manamah; access to the US Navy is also granted at Mina al-Sulman port and Muharraq airfield, and prepositioned equipment is sited at Shaikh Isa Air Base; Al Manamah hosts the Navy and Marine HQs of CENTCOM (NAVCENT and MARCENT).[56]
Kuwait	- *Ahmed Al Jaber Air Base* - *Ali Al Salem Air Base* - *Camp Arifjan* (CENTCOM Army HQ) - *Camp Buehring/Udairi Range* - *Camp Doha* - *Kuwait Naval Base*	Kuwait hosts thousands of troops and supporting personnel for Operation Iraqi Freedom at Camp Arifjan, Camp Buehring, Camp Doha, Kuwait Naval Base and Ali al-Salem Air Base; it has also granted the US Air Force access to Ahmed al Jaber Air Base; Camp Arifjan hosts the Army HQ of CENTCOM (ARCENT).

(Continued)

TABLE 1 (*Continued*)

Country	Facility[55]	Description
Oman	- *Khasab Air Base* - *Masirah Air Base* - *Muscat Port* - *Salalah Port* - *Seeb Air Base* - *Thumrait Air Base*	Oman accommodates prepositioned equipment and gives access rights to the US Air Force at Khasab, Masirah, Thumrait, and Seeb Air Bases, where it also hosts some US Air Force personnel; the ports of Muscat and Salalah also provide facilities.
Qatar	- *Al Udeid Air Base* (CENTCOM Air Force HQ) - *Camp As Sayliyah* (CENTCOM HQ and CENTCOM Special Operations HQ) - *Doha fuel storage location* - *Umm Said Port*	Qatar hosts CENTCOM's forward HQ in the Gulf at Camp As Saliyah, which also hosts the Special Operations HQ of CENTCOM (SOCENT), and prepositioned US Army materials; Al Udeid Air Base hosts the hub of the US Air Force operations in the Gulf and is the Air Force HQ of CENTCOM (CENTAF); Doha and Umm Said are also used for fuel storage and port facilities.
Saudi Arabia	- *Eskan Village Compound, Riyadh* - *Jeddah Air Base* - *King Abdul Aziz Air Base, Dhahran* - *King Abdul Aziz Naval Base, Jubail* - *King Khalid Air Force Base, Khamis-Mushayt* - *USMTM Compound, Tabuk*	After the launch of Operation Iraqi Freedom, the US withdrew and relocated to Qatar most of its Saudi-based forces by September 2003; US military personnel are still actively deployed, however, at various bases in Saudi Arabia directing the ongoing 'United States Military Training Mission to Saudi Arabia' (USMTM) initiated in 1953 in the aftermath of the Roosevelt-Aziz agreement on USS *Quincy* in 1945.
UAE	- *Al Dhafra Air Base* - *Jebel Ali Port* - *Fajairah International Airport*	The UAE hosts US supporting military personnel for Operations Iraqi Freedom and Operation Enduring Freedom at Al Dhafra Air Base; Jebel Ali port facilitates US ships resupplying Al Dhafra Air Base.

GEOPOLITICS AND BIOPOLITICS: THE 'TOXIC COMBINATION'

The sheer extent of the current US military forward presence in the GCC/Persian Gulf region is new in the American experience. For the first time, there now appears the contour of a continuous US ground presence, which has been further facilitated by the ongoing Iraq War and broader war on terror. And, of course, a host of US foreign policy strategists and security experts have enthusiastically scripted the geostrategic and geoeconomic opportunities attained under the rubric of the long war.[57] Indeed, as Asli Bâli and Aziz Rana underline, both the Republicans and Democrats "continue to take as given the centrality of pacification and global omnipresence for the promotion of American interests, despite the extent to which the experience of the last decade underscores the counter-productivity of these policies".[58] Today, as the long war continues unabated across the Middle East and Central Asia, the United States holds an unprecedented number of bases and access facilities across the most energy-rich region on earth. In Iraq, the US military has 45 major bases and well over 100 forward operating bases in total; in Afghanistan, it utilises over a dozen major base and airfield facilities; and, in addition, key bases and access facilities are maintained in Bahrain, Djibouti, Egypt, Jordan, the Kyrgyz Republic, Oman, Pakistan, Qatar, Saudi Arabia and the United Arab Emirates.[59] It has become clear too that the Pentagon is intent on establishing at least 14 "enduring bases" as the spoils of the Iraq War;[60] and there are various other 'projects of securitization' being planned for that reveal the 'long-term' vision for a permanent US ground presence across the Persian Gulf.[61] For the 'long war of securitization', the Pentagon's contingency plans for maintaining and extending its global ground presence – what it calls 'full-spectrum dominance' – can be read as a stark warning that on the US military's Zulu Time the sun never sets. But all of its 'land power' must still be secured and capacitated by extending the architecture and operation of the US military's biopolitical power on the ground. This is where biopolitics merges with geopolitics.

The US military's geostrategic forward presence in the Persian Gulf and elsewhere becomes only fully realised when its 'geopolitical operational capacity' is paralleled by a 'biopolitical operational capacity' on the ground. The latter must be enabled by a legal architecture allowing for, and governing, land access, troop circulation and conduct. This is the "toxic combination of geopolitics and biopolitics" that Michael Dillon has in mind when he observes the securitization practices of the war on terror.[62] For Dillon, the "geopolitics of security" revolve around the space of "territory", while the "biopolitics of security" revolve around the space of "population", yet both are indelibly intertwined.[63] Dillon's observation has been echoed by many. For Derek Gregory, for example, "biopolitics is not pursued outside the domain of sovereign power but is instead part of a protracted struggle over the right to claim, define and exercise sovereign power".[64] Of course, it could be argued that geo-politics has always centrally involved bio-politics too and that any recent drawing out of the multiple overlaps between the two simply reflects inadequate prior definitions of both classical and critical geopolitics.[65] In any case, what undoubtedly remains a challenge is the task of revealing and expounding how biopolitical strategies "relate to broader scale issues such as geopolitics and national economic and political policy, and vice versa".[66] It is this theorising of the complex relations between 'micro' and 'macro' scales of power that is key to Schlosser's call to "avoid dualistic notions of bio-political and sovereign power".[67]

Reflecting on the character of late modern war, Derek Gregory draws a useful distinction between the 'object-ontology' of traditional geopolitics, with its customary territorial concerns, and the 'event-ontology' of contemporary biopolitics, where battlefields are *composed* of events rather than objects" and biopolitical arguments are concerned with making interventionary military violence "appear to be intrinsically therapeutic".[68] The battle zones of the war on terror may well be "visibly and viscerally alive with death" (when, as Gregory observes, 'biopolitics' becomes 'necropolitics') but the US military are adept at navigating such potentially damning biopolitical registers in its dealings with the media. Long scripted geopolitical registers quickly become mobilised at press briefings, which reinstate "optical detachment" by reductively re-mapping battle spaces back into "an abstract geometry of points".[69] Discursively, the US military is certainly attuned to an expedient entwining of its biopolitical and geopolitical operations. And as I argue below, this discernible conflation of biopolitical and geopolitical strategies applies to the material securitization practices of the US military as well.

JURIDICAL WARFARE: FORWARD DEPLOYMENT BEYOND THE BATTLEFIELD

Nearly two centuries ago, Prussian military strategist, Carl von Clausewitz, observed how war is merely a "continuation of political commerce" by "other means".[70] Today, the lawfare of the US military is a continuation of war by legal means. Indeed, for US Deputy Judge Advocate General, Major General Charles Dunlap, it "has become a key aspect of modern war".[71] For Dunlap and his colleagues in the JAG corps, the law is a "force multiplier", as Harvard legal scholar, David Kennedy, explains: it "structures logistics, command, and control"; it "legitimates, and facilitates" violence; it "privileges killing"; it identifies legal "openings that can be made to seem persuasive", promissory, necessary and indeed therapeutic; and, of course, it is "a communication tool" too because defining the battlefield is not only a matter of "privileging killing", it is also a "rhetorical claim".[72] Viewed in this way, the law can be seen to in fact "contribute to the proliferation of violence rather than to its containment", as Eyal Weizman has instructively shown in the case of recent Israeli lawfare in Gaza.[73]

In the US wars in Iraq, Afghanistan and broader war on terror, the Department of Defense has actively sought to legalise its use of biopolitical violence against all those deemed a threat. Harvey Rishikof, the former Chair of the Department of National Security Strategy at the National War College in Washington, recently underlined 'juridical warfare' (his preferred designation over 'lawfare') as a pivotal "legal instrument" for insurgents in the asymmetric war on terror.[74] For Rishikof and his contemporaries, juridical warfare is always understood to mean the legal strategies of the weak 'against' the United States; it is never acknowledged as a legal strategy 'of' the United States. However, juridical warfare has been a proactive component of US military strategy overseas for some time, and since the September 11 attacks in New York and Washington in 2001, a renewed focus on juridical warfare has occurred, with the JAG Corps playing a central role in reforming, prioritising and mobilising the law as an active player in the war on terror.[75]

Deputy Judge Advocate General, Major General Charles Dunlap, recently outlined some of the key concerns facing his corps and the broader US military; foremost of which is the imposing of unnecessary legal restraints on forward-deployed military personnel.[76] For Dunlap, imposing legal restraints on the battlefield as a "matter of policy" merely "play[s] into the hands of those who would use [international law] to wage lawfare against us".[77] Dunlap's counter-strategy is simply "adhering to the rule of law", which "understands that sometimes the legitimate pursuit of military objectives will foreseeably – and inevitably – cause the death of noncombatants"; indeed, he implores that "this tenet of international law be thoroughly understood".[78] But 'the' rule of international law that Dunlap has in mind is merely a selective and suitably enabling set of malleable legal conventions that legitimate the unleashing of military violence.[79] As David Kennedy illuminates so brilliantly in *Of War and Law*:

> We need to remember what it means to say that compliance with international law "legitimates." It means, of course, that killing, maiming, humiliating, wounding people is legally privileged, authorized, permitted, and justified.[80]

The recent 'special issue on juridical warfare' in the US military's flagship journal, *Joint Force Quarterly*, brought together a range of leading judge advocates, specialists in military law, and former legal counsels to the Chairman of the Joint Chiefs of Staff. All contributions addressed the question of "which international conventions govern the confinement and interrogation of terrorists and how".[81] The use of the term 'terrorists' instead of suspects sets the tone for the ensuing debate: in an impatient defence of 'detention', Colonel James Terry bemoans the "limitations inherent in the Detainee Treatment Act of 2005 and the Military Commissions Act of 2006" (which he underlines only address detainees at the US Naval Base at Guantanamo) and asserts that "requirements inherent in the war on terror will likely warrant expansion of habeas corpus limitations";[82] considering 'rendition', Colonel Kevin Cieply asks the shocking question "is rendition simply recourse to the beast at a necessary time";[83] Colonel Peter Cullen argues for the necessity of the "role of targeted killing in the campaign against terror";[84] Commander Brian Hoyt contends that it is "time to re-examine U.S. policy on the [international criminal] court, and it should be done through a *strategic lens*";[85] while Colonel James Terry furnishes an additional concluding essay with the stunningly instructive title 'The International Criminal Court: A Concept Whose Time Has *Not* Come'.[86] These rather chilling commentaries attest to one central concern of the JAG Corps and the broader military-political executive at the Pentagon: that enemies must not be allowed to exploit "real, perceived, or even orchestrated incidents of law-of-war violations being employed as an unconventional means of confronting American military power".[87] And such thinking is entirely consistent with the defining *National Defense Strategy* of the Bush administration, which signalled the means to win the war on terror as follows: "We will defeat adversaries at the time, place, and in the manner of our choosing".[88]

If US warfare in the war on terror is evidently underscored by a 'manner of our choosing' preference – both at the Pentagon and in the battlefield – this in turn prompts an especially proactive 'juridical warfare' that must be simultaneously pursued to legally capacitate, regulate and maximise any, and all, military operations. The 2005 *National Defense Strategy* underlined the challenge thus:

> Many of the current legal arrangements that govern overseas posture date from an earlier era. Today, challenges are more diverse and complex, our prospective contingencies are more widely dispersed, and our international partners are more numerous. International agreements relevant to our posture must reflect these circumstances and support greater operational flexibility.[89]

It went on to underline its consequent key juridical tactic and what I argue is a critical weapon in the US military-legal arsenal in the war on terror: the securing of 'Status of Forces Agreements' – to "provide legal protections" against "transfers of U.S. personnel to the International Criminal Court".[90]

A 'Status of Forces Agreement' (SOFA) is "an agreement that defines the legal position of a visiting military force deployed in the territory of a friendly state".[91] Each of the SOFAs secured by the US military with its host countries in the war on terror are classified. However, the vital components of any SOFA serve to primarily define the legal designation of military personnel to negate accountability to both national and international law.[92] And this is all done a priori, of course; SOFAs anticipate the 'aleatory' and plan for the 'evental'. Their essential purpose, indeed, echoes resoundingly Foucault's theorisation of the 'biopolitical apparatus of security': to "fabricate, organize, and plan a milieu".[93] In practical terms, the critical importance of a SOFA is well summed up by US Air Force JAG Lieutenant Colonel Jeffrey Walker in his blunt assertion that "without a SOFA, you're just a group of heavily armed tourists".[94] The key role of Military JAGs, then, is to legally condition and safeguard the deployment of US forces overseas. As one Assistant Staff

Judge Advocate at US Southern Command set out in considerable detail in *The Army Lawyer* in 2000, the basic JAG responsibility is to "keep military personnel from going to jail for doing the right thing";[95] or more specifically for doing what is legally enabled.

For CENTCOM, securing SOFAs with various Middle Eastern countries has been a critical element of theatre strategy since the early 1980s.[96] Since the war on terror began, however, the command has broadened negotiations with various countries in its AOR to formalise military ties. Uzbekistan is one such country. Just three weeks after the September 11 attacks, "concerted negotiations involving teams of CENTCOM JAGs and State Department lawyers" culminated in the Uzbeks signing a crucial SOFA for the US military, which permitted US forces access to Uzbekistani territory and airspace in preparation for the then imminent attack on Afghanistan.[97] This was the first action in the war on terror – on the legal battlefield. The US has since signed defence pacts and SOFA agreements with various other allies in its long war against terrorism, including Kuwait, Bahrain, Qatar and Djibouti.[98]

The most important element of any SOFA is the establishment of the legal jurisdiction within which foreign armed personnel operate in host countries. SOFAs, in effect, define the "legal status of the foreign troops" by "setting forth the rights and responsibilities between the basing and hosting power with regard to such matters as criminal and civil jurisdiction".[99] In the war on terror, the Bush administration consistently signalled its intentions to exempt US forces from accountability to the jurisdiction of both host governments and international law.[100] On the national level, this resulted in the protection of US personnel, such as Airman Zachary Hatfield who fatally shot a Kyrgyz civilian at a checkpoint at Manas Air Base in December 2006, from local prosecution; Hatfield was sent home to the United States and hence out of the jurisdiction of the Kyrgyz Republic in March 2007.[101] On the international level, the unilateral Bush agenda routinely resulted in the bypassing and refusal of international legal jurisdiction in the governance of US military action. Any checking of US violations of international law was vitriolically resisted. In 2005, for example, just a few days after DePaul Professor of International Law, M. Cherif Bassiouni, released a UN report criticising the US military for committing human rights abuses in Afghanistan, intense US pressure saw him removed as 'UN independent expert on human rights' in the country.[102]

Under the rubric of 'Transforming the US Global Defense Posture', initiated in 2003, the Department of Defense factored the careful negotiation of SOFAs that eschew the jurisdiction of international law centrally into its long-term military strategy overseas. Speaking at the Center for Strategic and International Studies in Washington in December 2003, the then Under Secretary of Defense for Policy, Douglas Feith, set out the "transformation" in "longer-term thinking about U.S. defense strategy", which revolved around developing "rapidly deployable capabilities" worldwide – capabilities that rely upon legal-biopolitical technologies of power.[103] Feith outlined the necessary lawfare required to facilitate such a grandiose strategy:

> For this deployability concept to work, US forces must be able to move smoothly into, through, and out of host nations, which puts a premium on establishing legal and support arrangements. . . . We are negotiating or planning to negotiate with many countries legal protections for US personnel, through Status of Forces Agreements and agreements (known as Article 98 agreements) limiting the jurisdiction of the International Criminal Court with respect to our forces' activities.[104]

Feith intimated too the financial trade-off for countries willing to participate: "We are putting in place so-called cross-servicing agreements so that we can rapidly reimburse countries for support they provide to our military operations".[105] Like all warfare, 'juridical warfare' pays someone.

The rapid deployability concept was officially codified with the publication and report to Congress of the *Global Defense Posture Review* in 2004.[106] Therein, "bilateral and multilateral legal arrangements" are underlined as critical components of "global defense posture", which allow for the "necessary flexibility and freedom of action to meet 21st-century security challenges".[107] Defense Under Secretary Feith's previously announced design to bypass international law and specifically the International Criminal Court in future negotiations of access agreements is also explicitly signalled.[108] The 2005 *National Defense Strategy* further reinforced the US unilateral position; its bold warning that "our strength as a nation will continue to be challenged by those who employ a strategy of the weak using international fora, judicial processes, and terrorism" representing a stark illustration of the Bush administration's contempt for international institutions and international law.[109] If we were in any doubt as to the Bush administration's anticipated 'milieu' for Iraq, a situation the Obama administration has inherited, we got a clear confirmation of it in Patrick Cockburn's outlining of leaked details of a "secret deal" negotiated in Baghdad in the summer of 2008 that would "perpetuate the American military occupation of Iraq indefinitely, regardless of the outcome of the US presidential election": "Bush wants 50 military bases, control of Iraqi airspace and legal immunity for all American soldiers and contractors".[110] This is the US military's securitization endgame overseas: territorially secured land power and legally enabled biopower.

SECURITY, NOT LIBERTY: THE 'PERMANENT EMERGENCY' OF THE SECURITY SOCIETY

The US military's evident disdain for international law, indifference to the pain of 'Others' and endless justifying of its actions via the language of 'emergency' have prompted various authors to reflect on Giorgio Agamben's work, in particular, on bare life and the state of exception in accounting for the functioning of US sovereign power in the contemporary world.[111] Claudio Minca, for example, has used Agamben to attempt to lay bare US military power in the spaces of exception of the global war on terror; for Minca, "it is precisely the absence of a theory of space able to inscribe the spatialisation of exception that allows, today, such an enormous, unthinkable range of action to sovereign decision".[112] This critique speaks especially to the excessive sovereign violence of our times, all perpetrated in the name of a global war on terror.[113] Minca's argument is that geography as a discipline has failed to geo-graph and theorise the spatialisation of the 'pure' sovereign violence of legitimated geopolitical action overseas. He uses the notion of the camp to outline the spatial manifestation and endgame of a new global biopolitical 'nomos' that has unprecedented power to except bare life.[114]

In the 'biopolitical nomos' of camps and prisons in the Middle East and elsewhere, managing detainees is an important element of the US military project. As CENTCOM Commander General John Abizaid made clear to the Senate Armed Services Committee in 2006, "An essential part of our combat operations in both Iraq and Afghanistan entails the need to detain enemy combatants and terrorists".[115] However, it is a mistake to characterise as 'exceptional' the US military's broader biopolitical project in the war on terror. Both Minca's and Agamben's emphasis on the notion of 'exception' is most convincing when elucidating how the US military has dealt with the 'threat' of enemy combatants, rather than how it has planned for, legally securitized and enacted, its 'own' aggression against them. It does not account for the proactive juridical warfare of the US military in its forward deployment throughout the globe, which rigorously secures classified SOFAs with host nations and protects its armed personnel from transfer to the International Criminal Court. Far from designating a 'space of exception', the US does this to establish normative parameters in its exercise of legally sanctioned military violence and to maximise its 'operational capacities of securitization'.

A bigger question, of course, is what the US military practices of lawfare and juridical securitization say about our contemporary moment. Are they essentially 'exceptional' in character, prompted by the so-called exceptional character of global terrorism today? Are they therefore enacted in 'spaces of exceptions' or are they, in fact, simply contemporary examples of Foucault's 'spaces of security' that are neither exceptional nor indeed a departure from, or perversion of, liberal democracy? As Mark Neocleous so aptly puts it, has the "liberal project of 'liberty'" not always been, in fact, a "project of security"?[116] This 'project of security' has long invoked a powerful political *dispositif* of 'executive powers', typically registered as 'emergency powers', but, as Neocleous makes clear, of the permanent kind.[117] For Neocleous, the pursuit of 'security' – and more specifically 'capitalist security' – marked the very emergence of liberal democracies, and continues to frame our contemporary world. In the West at least, that world may be endlessly registered as a liberal democracy defined by the 'rule of law', but, as Neocleous reminds us, the assumption that the law, decoupled from politics, acts as the ultimate safeguard of democracy is simply false – a key point affirmed by considering the US military's extensive waging of liberal lawfare. As David Kennedy observes, the military lawyer who "carries the briefcase of rules and restrictions" has long been replaced by the lawyer who "participate[s] in discussions of strategy and tactics".[118]

The US military's liberal lawfare reveals how the rule of law is simply another securitization tactic in liberalism's 'pursuit of security'; a pursuit that paradoxically eliminates fundamental rights and freedoms in the 'name of security'.[119] This is a 'liberalism' defined by what Michael Dillon and Julian Reid see as a commitment to waging 'biopolitical war' for the securitization of life – 'killing to make live'.[120] And for Mark Neocleous, (neo)liberalism's fetishisation of 'security' – as both a discourse and a technique of government – has resulted in a world defined by anti-democratic technologies of power.[121] In the case of the US military's forward deployment on the frontiers of the war on terror – and its juridical tactics to secure biopolitical power threat – this has been made possible by constant reference to a neoliberal 'project of security' registered in a language of 'endless emergency' to 'secure' the geopolitical and geoeconomic goals of US foreign policy.[122] The US military's continuous and indeed growing military footprint in the Middle East and elsewhere can be read as a 'permanent emergency',[123] the new 'normal' in which geopolitical military interventionism and its concomitant biopolitical technologies of power are necessitated by the perennial political economic 'need' to securitize volatility and threat.

CONCLUSION: ENABLING BIOPOLITICAL POWER IN THE AGE OF SECURITIZATION

> Law and force flow into one another. We make war in the shadow of law, and law in the shadow of force.
> — David Kennedy, *Of War and Law* [124]

Can a focus on lawfare and biopolitics help us to critique our contemporary moment's proliferation of practices of securitization – practices that appear to be primarily concerned with coding, quantifying, governing and anticipating life itself? In the context of the US military's war on terror, I have argued above that it can. If, as David Kennedy points out, the "emergence of a global economic and commercial order has amplified the role of background legal regulations as the strategic terrain for transnational activities of all sorts", this also includes, of course, 'warfare'; and for some time, the US military has recognised the "opportunities for creative strategy" made possible by proactively waging lawfare beyond the battlefield.[125] As Walter Benjamin observed nearly a century ago, at the very heart of military violence is a "lawmaking character".[126] And it is this 'lawmaking character' that is integral to the biopolitical technologies of power that secure US geopolitics in our contemporary moment. US lawfare focuses "the attention of the world on this or that excess" whilst simultaneously arming "the most heinous human suffering in legal privilege", redefining horrific violence as "collateral damage, self-defense, proportionality, or necessity".[127] It involves a mobilisation of the law that is precisely channelled towards "evasion", securing classified Status of Forces Agreements and "offering at once the experience of safe ethical distance and careful pragmatic assessment, while parcelling out responsibility, attributing it, denying it – even sometimes embracing it – as a tactic of statecraft and war".[128]

Since the inception of the war on terror, the US military has waged incessant lawfare to legally securitize, regulate and empower its 'operational capacities' in its multiples 'spaces of security' across the globe – whether that be at a US base in the Kyrgyz Republic or in combat in Iraq. I have sought to highlight here these tactics by demonstrating how the execution of US geopolitics relies upon a proactive legal-biopolitical securitization of US troops at the frontiers of the American 'leasehold empire'. For the US military, legal-biopolitical apparatuses of security enable its geopolitical and geoeconomic projects of security on the ground; they plan for and legally condition the 'milieux' of military commanders; and in so doing they render operational the pivotal spaces of overseas intervention of contemporary US national security conceived in terms of 'global governmentality'.[129] In the US global war on terror, it is lawfare that facilitates what Foucault calls the "biopolitics of security" – when life itself becomes the "object of security".[130] For the US military, this involves the eliminating of threats to 'life', the creating of operational capabilities to 'make live' and the anticipating and management of life's uncertain 'future'.

Some of the most key contributions across the social sciences and humanities in recent years have divulged how discourses of 'security', 'precarity' and 'risk' function centrally in the governing *dispositifs* of our contemporary world.[131] In a society of (in)security, such discourses have a profound power to invoke danger as "requiring extraordinary action".[132] In the ongoing war on terror, registers of emergency play pivotal roles in the justification of military securitization strategies, where 'risk', it seems, has become permanently binded to 'securitization'. As Claudia Aradau and Rens Van Munster point out, the "perspective of risk management" seductively effects practices of military securitization to be seen as necessary, legitimate and indeed therapeutic.[133] US tactics of liberal lawfare in the long war – the conditioning of the battlefield, the sanctioning of the privilege of violence, the regulating of the conduct of troops, the interpreting, negating and utilising of international law, and the securing of SOFAs – are vital security *dispositifs* of a broader 'risk-securitization' strategy involving the deployment of liberal technologies of biopower to "manage dangerous irruptions in the future".[134] It may well be fought *beyond* the battlefield in "a war of the pentagon rather than a war of the spear",[135] but it is lawfare that ultimately enables the 'toxic combination' of US geopolitics and biopolitics defining the current age of securitization.

ACKNOWLEDGEMENTS

My thanks to Simon Dalby, Derek Gregory, Gerry Kearns, David Nally, Neil Smith and two anonymous referees for their insightful comments, and to Jochen Albrecht for drawing Figures 1 and 2. I want to also gratefully acknowledge the support of the Irish Research Council for the Humanities and Social Sciences while I was a fellow at CUNY Graduate Centre in 2007/2008, during which time I carried out the research for this paper.

NOTES

1. See, for example: D. Gregory, 'The Black Flag: Guantánamo Bay and the Space of Exception', *Geografiska Annaler B* 88/4 (2006) pp. 405–427; S. Reid-Henry, 'Exceptional Sovereignty? Guantanamo Bay and the Re-Colonial Present', *Antipode* 39/4 (2007) pp. 627–648; G. Kearns, 'The Geography of Terror', *Political Geography* 27/3 (2008) pp. 360–364; A. D. Barder, 'Power, Violence and Torture: Making Sense of Insurgency and Legitimacy Crises in Past and Present Wars of Attrition', in F. Debrix and M. Lacy (eds.), *The Geopolitics of American Insecurity: Terror, Power and Foreign Policy* (New York: Routledge 2009) pp. 54–70; and A. Macklin, 'Transjudicial Conversations about Security and Human Rights', in M. B. Salter (ed.), *Mapping Transatlantic Security Relations The EU, Canada and the War on Terror* (New York: Routledge 2010) pp. 212–235.

2. Both strategies work in concert with each other of course, with various locations acting as both 'base' and 'prison'.

3. US Department of Defense, *The National Defense Strategy of the United States of America 2005* (Washington, DC: Department of Defense 2005) p. 9.

4. H. Rishikof, 'Juridical Warfare: The Neglected Legal Instrument', *Joint Force Quarterly* 48 (2008) p. 11.

5. W. Benjamin, *Reflections: Essays, Aphorisms, Autobiographical Writings*, trans. by E. Jephcott (New York: Harcourt Brace Jovanovich 1978) p. 283.

6. M. Foucault, *Security, Territory, Population: Lectures at the Collège de France, 1977–1978*, trans. by G. Burchell (Basingstoke: Palgrave Macmillan 2007) p. 10.

7. L. Amoore, 'Biometric Borders: Governing Mobilities in the War on Terror', *Political Geography* 25/3 (2006) pp. 336–351; M. Dillon, 'Governing Through Contingency: The Security of Biopolitical Governance', *Political Geography* 26/1 (2007) pp. 41–47; J. Donzelot, 'Michel Foucault and Liberal Intelligence', *Economy and Society* 37/1 (2008) pp. 115–134; J. Reid, *The Biopolitics of the War on Terror* (Manchester: Manchester University Press 2009).

8. J. W. Crampton and S. Elden (eds.), *Space, Knowledge and Power: Foucault and Geography* (Farnham, Surrey: Ashgate 2007); K. Schlosser, 'Bio-Political Geographies', *Geography Compass* 2/5 (2008) pp. 1621–1634; M. Coleman and K. Grove, 'Biopolitics, Biopower, and the Return of Sovereignty', *Environment and Planning D: Society and Space* 27/3 (2009) pp. 489–507.

9. See, for example, M. Dillon and A. W. Neal (eds.), *Foucault on Politics, Security and War* (Basingstoke: Palgrave Macmillan 2008).

10. "Problems of space" and the "treatment of space" were focal to his theorising of sovereignty, discipline and security, and their complex relational "multiplicities"; see Foucault (note 6) pp. 11–23.

11. D. Gregory, 'Seeing Red: Baghdad and the Event-Ful City', *Political Geography* 29/5 (2010) pp. 266–279.

12. Such transitions were certainly taking place earlier in the Early Modern Period (particularly with the onset of the expansion of the major European centralised states of England and Spain to their various new colonial populations); in Foucault's extensive canon of work, however, colonial societies have been largely overlooked. Foucault's thinking and lines of political inquiry were not, of course, without various other lacunae and unresolved problematics, as Michael Dillon and Andrew Neal have recently pointed out; see Dillon and Neal (note 9) Introduction.

13. Foucault (note 6) pp. 10, 11.

14. Ibid., pp. 8–23; Donzelot (note 7) p. 118.

15. Foucault (note 6) p. 70.

16. Ibid., pp. 116, 117, 120.

17. Ibid., p. 108.

18. Ibid., pp. 109, 110.

19. J. Morrissey, 'Closing the Neoliberal Gap: Risk and Regulation in the Long War of Securitization', *Antipode* 43/4 (2011) doi:10.1111/j.1467-8330.2010.00823.x. In theorising the securitization strategy of the contemporary US military, I have drawn especially upon the Copenhagen school of security studies use of the term 'securitization', understood as a powerful discourse through which 'security risks' come to legitimise the use of 'emergency powers' and 'apparatuses of security'; see, for example, B. Buzan, O. Wæver, and P. de Wilde P, *Security: A New Framework for Analysis* (Boulder, CO: Lynne Rienner 1997); and M. Williams, *Culture and Security: Symbolic Power and the Politics of International Security* (London: Routledge 2007). See also S. Dalby, *Security and Environmental Change* (Cambridge: Polity 2009) pp. 46–49.

20. I have recently explored this powerful assemblage of what are, in effect, classical geopolitical knowledges, advanced by what can be termed Washington's 'military-strategic studies complex': J. Morrissey, 'Architects of Empire: The Military-Strategic Studies Complex and the Scripting of US National Security', *Antipode* 43/2 (2011) pp. 435–470.

21. Foucault (note 6) pp. 20–21.

22. The US military's six regional commands are: US Northern Command (NORTHCOM); US Southern Command (SOUTHCOM); US Pacific Command (PACOM); US European Command (EUCOM); US Central Command (CENTCOM); and US Africa Command (AFRICOM; the most recently established command, attaining 'full operational capabilities' in October 2008).

23. Foucault (note 6) p. 20.

24. Source: US Department of Defense, 'U.S. Africa Command', briefing slide (7 Feb. 2007), available at <http://www.defenselink.mil/dodcmsshare/briefingslide%5C295%5C070207-D-6570C-001.pdf>, accessed 9 Feb. 2008.

25. Foucault (note 6) p. 70.

26. Ibid., pp. 20, 21.

27. Ibid.

28. Ibid., p. 21.

29. Ibid.

30. Dillon, 'Governing Through Contingency' (note 7) p. 46.

31. M. Dillon, 'Governing Terror: The State of Emergency of Biopolitical Emergence', *International Political Sociology* 1/1 (2007) p. 12.

32. M. Hannah, 'Torture and the Ticking Bomb: The War on Terrorism as a Geographical Imagination of Power/Knowledge', *Annals of the Association of American Geographers* 96/3 (2006) p. 622.

33. R. Martin, *An Empire of Indifference: American War and the Financial Logic of Risk Management* (Durham: Duke University Press 2007).

34. N. Klein, *The Shock Doctrine: The Rise of Disaster Capitalis*m (New York: Metropolitan Books 2007).

35. R. Grusin, 'Premediation', *Criticism* 46/1 (2004) p. 19; M. de Goede, 'Beyond Risk: Premediation and the Post-9/11 Security Imagination', *Security Dialogue* 39/2–3 (2008) p. 158.

36. Foucault (note 6) p. 33.

37. Ibid., pp. 42–43.

38. D. Nally, 'The Biopolitics of Food Provisioning', *Transactions of the Institute of British Geographers* (forthcoming).

39. Dillon, 'Governing Terror' (note 31) p. 9.

40. A. Krepinevich and R. O. Work, *A New Global Defense Posture for the Second Transoceanic Era* (Washington, DC: Center for Strategic and Budgetary Assessments 2007). Cf. Morrissey, 'Architects of Empire' (note 20).

41. US Department of Defense, *Active Duty Military Personnel Strengths by Regional Area and by Country* (31 Dec. 2009), available at <http://siadapp.dmdc.osd.mil/personnel/MILITARY/history/hst0912.pdf>, accessed 10 June 2010.

42. See J. Morrissey, 'The Geoeconomic Pivot of the Global War on Terror: US Central Command and the War in Iraq', in D. Ryan and P. Kiely (eds.), *America and Iraq: Policy-Making, Intervention and Regional Politics* (New York: Routledge 2009) pp. 103–122.

43. US Department of Defense, *Authorization for Appropriations for Fiscal Year 1981: Hearings of the Senate Armed Services Committee* (Washington, DC: Senate Armed Services Committee 1980) p. 484.

44. US Congressional Budget Office, *Rapid Deployment Forces: Policy and Budgetary Implications* (Washington, DC: US Congressional Budget Office 1983) p. 38. In the contemporary world, prepositioning continues to be a key tool in advancing US geostrategic capabilities overseas – its current aim is to "keep the capacity to quickly fill out up to eight ground combat brigades with equipment stored overseas"; see M. O'Hanlon, *Unfinished Business: U.S. Overseas Military Presence in the 21st Century* (Washington, DC: Center for a New American Security 2008) p. 14.

45. J. Record, 'The Rapid Deployment Force: Problems, Constraints, and Needs', *Annals of the American Academy of Political and Social Science* 457 (1981) pp. 109, 112.

46. Ibid., p. 112.

47. J. M. Epstein, 'Soviet Vulnerabilities in Iran and the RDF Deterrent', *International Security* 6/2 (1981) p. 127. See also: T. L. McNaugher, *Arms and Oil: U.S. Strategy and the Persian Gulf* (Washington, DC: Brookings Institution 1985); and J. M. Epstein, *Strategy and Force Planning: The Case of the Persian Gulf* (Washington, DC: Brookings Institution 1987).

48. Further *Bright Star* exercises in 1981, 1983, 1985 and 1987, saw up to 10,000 US troops at a time training in the desert with troops from Egypt, Jordan, Oman, Somalia and Sudan. From *Bright Star* 1983 onwards, Egypt emerged as the most committed military partner in the region in the 1980s, and therefore its role in "American preparations for the defence of Southwest Asia" became a "key component"; see D. Gold, *America, the Gulf and Israel: CENTCOM (Central Command) and Emerging US Regional Security Policies in the Middle East* (Jerusalem: Jaffee Center for Strategic Studies 1988) p. 41. With the Department of Defense focused on deterring an imminent Soviet attack in the region (that never of course materialised), the Egyptian army's use of Soviet military equipment made it an excellent training partner for the first CENTCOM forces deployed in the Middle East. Egypt's importance to US grand strategy in the Middle East continues to this day; it was the only African country retained in CENTCOM's AOR upon the initiation of the new Africa unified command, AFRICOM, in 2008.

49. J. Record, *The Rapid Deployment Force and U.S. Military Intervention in the Persian Gulf* (Washington, DC: Institute for Foreign Policy Analysis 1981) p. 58. In Oman, US access was given at Muscat airfield and port, Seeb airfield and port, Masirah airfield, Salalah airfield and Thumrait airfield; Somalia granted use permissions at Berbera airfield and port and Mogadishu airfield and port; while Kenya approved access at Mombasa port, Nairobi airfield and Nanyuki airfield; see ibid., p. 59.

50. See Krepinevich and Work (note 40) pp. 149–150; see also D. Priest, *The Mission: Waging War and Keeping Peace with America's Military* (New York: W.W. Norton and Co. 2003) pp. 78–98.

51. S. G. Hajjar, *U.S. Military Presence in the Gulf: Challenges and Prospects* (Carlisle, PA: Strategic Studies Institute, U.S. Army War College 2002) p. 20.

52. Overseas Basing Commission, *Commission on Review of the Overseas Military Facility Structure of the United States Report to the President and Congress* (Arlington, VA: Commission on Review of Overseas Military Facility Structure of the United States 2005) pp. iii, 6–7.

53. Ibid., p. G5.

54. US CENTCOM Director of Logistics Major General Brian Geehan, 'Current Operations in the Long War', National Defense Transportation Association Forum and Expo, Memphis, Tennessee, 23–27 Sep. 2006.

55. Various sources attribute a far greater number of bases and access facilities to each GCC country; for example, Global Security (globalsecurity.org) puts the number of US major bases in Kuwait

at 16. However, this table is conservatively based on official US government data, compiled from the following sources: Overseas Basing Commission (note 52); K. Katzman, *The Persian Gulf States: Issues for U.S. Policy, 2006* (Washington, DC: Congressional Research Service Report for Congress 2006); US Department of Defense, *Base Structure Report Fiscal Year 2007 Baseline* (Washington, DC: Office of the Deputy Under Secretary of Defense (Installations and Environment) 2007); and United States Military Training Mission to Saudi Arabia, 'USMTM Welcome Package' (13 June 2010), available at <http://www.usmtm.sppn.af.mil/newFRG2/Documents/Newcomer-booklet.pdf>, accessed 12 July 2010.

56. Prior to the current Iraq war, NAVCENT was the only component command of CENTCOM permanently located in its AOR.

57. See, for example, B. A. Thayer, *The Pax Americana and the Middle East: U.S. Grand Strategic Interests in the Region after September 11* (Ramat Gan, Israel: Bar-Ilan University, The Begin-Sadat Center for Strategic Studies 2003) and, more recently, R. D. Kaplan, 'The Revenge of Geography', *Foreign Policy* (May/June 2009), available at <http://www.foreignpolicy.com/story/cms.php?story_id=4862>, accessed 6 May 2009.

58. A. Bâli and A. Rana, 'American Overreach: Strategic Interests and Millennial Ambitions in the Middle East', *Geopolitics* 15/2 (2010) p. 233.

59. See: Overseas Basing Commission (note 52) Appendix G; Katzman (note 55) pp. 7–11; and O'Hanlon (note 44) pp. 28–32.

60. J. Gerson, "Enduring' U.S. Bases in Iraq: Monopolizing the Middle East Prize', *Common Dreams News Center* (19 March 2007), available at <http://www.commondreams.org/views07/0319-26.htm>, accessed 12 Feb. 2008. In a showing of its renewed confidence for the success of Operation Iraqi Freedom, the US also announced plans of a further military base just four miles short of the Iraq-Iran border; see BBC News, 'US Plans Base on Iraq-Iran Border' (10 Sep. 2007), available at <http://news.bbc.co.uk/2/hi/middle_east/6987306.stm>, accessed 25 Jan. 2008. Furthermore, the figure of fourteen 'enduring bases' appears to be quite a conservative estimate, as Patrick Cockburn recently revealed in *The Independent*; Cockburn details leaked press reports that put the number of long-term bases being planned for by the US military in Iraq at fifty; see P. Cockburn, 'Revealed: Secret Plan to Keep Iraq Under US Control', *The Independent*, 5 June 2008.

61. In March 2006, for example, the Pentagon announced a ten-year plan for 'deep storage' of weapons, munitions, equipment and supplies in six countries in the CENTCOM AOR, and by 2016 the tonnage of air munitions alone, stored at sites outside Iraq, will be double the 2006 levels; see W. M. Arkins, 'U.S. Plans New Bases in Iraq', Early Warning Blog, *The Washington Post* (22 March 2006), available at <http://blog.washingtonpost.com/earlywarning/2006/03/us_plans_new_bases_in_the_midd.html>, accessed 25 Jan. 2008.

62. Dillon, 'Governing Terror' (note 31) p. 9.

63. Dillon, 'Governing Through Contingency' (note 7) p. 46.

64. D. Gregory, 'The Biopolitics of Baghdad: Counterinsurgency and the Counter-City', *Human Geography* 1/1 (2008) p. 19.

65. This is part of Bruce Braun's argument about why it is "necessary to trace the ways that biopolitics has merged with geopolitics"; see B. Braun, 'Biopolitics and the Molecularization of Life', *Cultural Geographies* 14/1 (2007) p. 8. For Agamben, "Western politics", at its very heart, equated to a form of "bio-politics from the very beginning"; see G. Agamben, *Homo Sacer: Sovereign Power & Bare Life* (Stanford: Stanford University Press 1998) p. 181.

66. Schlosser (note 8) p. 1624. As Steve Legg argues, the very concept of governmentality is "trans-scalar"; see S. Legg, 'Foucault's Population Geographies: Classification, Biopolitics and Governmental Spaces', *Population, Space and Place* 11/3 (2005) p. 142.

67. Schlosser (note 8) p. 1631.

68. Gregory, 'Seeing Red' (note 11).

69. Ibid.

70. C. von Clausewitz, *On War*, trans. by J. J. Graham and Introduction by J. W. Honig. New York: Barnes and Noble 2004) p. 17.

71. C. J. Dunlap, 'Lawfare Amid Warfare', Op-Ed, *The Washington Times* (3 Aug. 2007), available at <http://www.washingtontimes.com/news/2007/aug/3/lawfare-amid-warfare>, accessed 10 July 2008.

72. D. Kennedy, *Of War and Law* (Princeton: Princeton University Press 2006) pp. 37, 121, 122, 148, 167.

73. E. Weizman, 'Lawfare in Gaza: Legislative Attack', *openDemocracy* (1 March 2009), available at <http://www.opendemocracy.net/article/legislative-attack>, accessed 10 July 2009. Weizman soberly concludes that "rather than moderation and restraint, the violence and destruction of Gaza might be the true face of international law" and urges anyone "concerned with the interests and rights of people affected by war" to employ a "double, even paradoxical strategy" that "uses international humanitarian law, while highlighting the dangers implied in it and challenging its truth claims and thus also the basis of its authority" (ibid.). Weizman's argument here echoes strongly David Kennedy's concerns about the recent "emergence of a powerful legal vocabulary for articulating humanitarian ethics in the context of war" (Kennedy (note 72) p. 8). Kennedy asks how do we react when we "find the humanist vocabulary of international law mobilised by the military as a strategic asset" or when the military "legally conditions the battlefield" by "informing the public that they *are entitled* to kill civilians" or when "our political leadership justifies warfare in the language of human rights" (ibid.).

74. Rishikof (note 4) p. 11. The term 'lawfare' was one of several alternative war-making concepts outlined by two Chinese People's Liberation Army officers in 1999: Q. Liang and W. Xiangsui, *Unrestricted Warfare* (Beijing: PLA Literature and Arts Publishing House 1999).

75. The Department of Defense's Defense Legal Services Agency received a staff increase of 100 FTEs from FY 2005 to FY 2006; see US Office of the Secretary of Defense, *Fiscal Year (FY) 2007 Budget Estimates, Volume II, Defense Wide Data Book* (Washington, DC: Office of the Secretary of Defense 2006) p. 26. The proliferation of 'lawfare' of course does prompt the question posed by the Comaroffs regarding what does the fixation on legalities in times of disorder really mean. Does the contemporary fetishisation of the law simply mask (as throughout colonial history) the contemporary American empire's relative impotence? See J. Comaroff and J. L. Comaroff, 'Law and Disorder in the Postcolony: An Introduction', in J. Comaroff and J. L. Comaroff (eds.), *Law and Disorder in the Postcolony* (Chicago: University of Chicago Press 2006) pp. 1–56.

76. Dunlap (note 71).

77. Ibid.

78. Ibid.

79. As the war on terror began, the international Geneva Conventions, for example, were quickly identified as restrictive for the 'new paradigm' of modern warfare. In January 2002, Attorney General Alberto Gonzales, as White House Chief Legal Counsel, advised President Bush that the war on terrorism "render[ed] obsolete Geneva's strict limitations on questioning of enemy prisoners and renders quaint some of its provisions"; see Center for American Progress, *Memorandum on the Geneva Conventions* (18 May 2004), available at <http://www.americanprogress.org/issues/kfiles/b79532.html>, accessed 1 July 2008.

80. Kennedy (note 72) p. 8.

81. Rishikof (note 4) p. 12. The 2007 conference, New Battlefields, Old Laws, held in Washington and jointly organised by the Institute for National Security and Counterterrorism (INSCT) at Syracuse University and Institute for Counter Terrorism (ICT) in Herzliya, Israel, addressed similar issues. The joint concern of the INSCT and ICT is that "recent conflicts underscore the continuing shortcomings of international law and policy in responding to asymmetric warfare mounted by non-state terrorist groups in the 21st century" (<http://insct.syr.edu/Battlefields/overview.htm>, accessed 10 July 2008). The conference and goals of both institutes has stirred considerable controversy; see, for example, Howard Friel's article 'Changing the Laws of War: Conference Seeks to Legitimize Civilian Casualties' in the *National Catholic Reporter* (5 Oct. 2007), available at <http://ncronline.org/index1005.htm>, accessed 10 July 2008.

82. J. P. Terry, 'Habeas Corpus and the Detention of Enemy Combatants in the War on Terror', *Joint Force Quarterly* 48 (2008) p. 18.

83. K. M. Cieply, 'Rendition: The Beast and the Man', *Joint Force Quarterly* 48 (2008) p. 20.

84. P. M. Cullen, 'The Role of Targeted Killing in the Campaign Against Terror', *Joint Force Quarterly* 48 (2008) p. 22.

85. B. A. Hoyt, 'Rethinking the U.S. Policy on the International Criminal Court', *Joint Force Quarterly* 48 (2008) p. 34 (my emphasis).

86. J. P. Terry, 'The International Criminal Court: A Concept Whose Time Has *Not* Come', *Joint Force Quarterly* 48 (2008) pp. 36–40.

87. Dunlap (note 71).

88. US Department of Defense, *National Defense Strategy* (note 3) p. 9.

89. Ibid., p. 19.

90. Ibid., p. 20.

91. US Department of Defense, *Dictionary of Military and Associated Terms, Joint Publication 1–02* (Washington, DC: Department of Defense 2007) p. 512. This definition derives from various Department of Defense (DoD) directives and instructions dating back to the late 1970s: DoD, Instruction 2050.1, 6 July 1977; DoD Directive 5525.1, 7 Aug. 1979; DoD Directive 5530.3, 6 Dec. 1979; and DoD Directive 3310.1, 22 Oct. 1982; see also the recently amended Chairman of the Joint Chiefs of Staff Instruction 2300.01D, *International Agreements* (2007), available at <http://www.dtic.mil/cjcs_directives/cdata/unlimit/2300_01.pdf>, accessed 7 July 2008.

92. SOFAs also typically serve to legally securitize facilities access and lethal and non-lethal US equipment prepositioning; they commonly set out too commitments to joint training exercises; and finally a key component secures arms sales; see Katzman (note 55) p. 7.

93. Foucault (note 6) p. 21.

94. Quoted in: W. C. Smith, 'Lawyers at War', *American Bar Association Journal* 89/2 (2003) p. 14. Smith outlines the multiple imports of a SOFA as follows: "SOFAs govern mission-critical and mundane matters including base access and security; provision or purchase of food, fuel, electricity and other supplies; taxes, visas and customs regulations; and criminal jurisdiction over U.S. personnel" (ibid.).

95. W. A. Stafford, 'How to Keep Military Personnel from Going to Jail for Doing the Right Thing: Jurisdiction, ROE and the Rules of Deadly Force', *The Army Lawyer*, Nov. 2000, p. 1. *The Army Lawyer*, published by The Judge Advocate General's School of the US Army at Charlottesville, Virginia, has been in publication since 1971; its stated purpose is "to provide practical, how-to-do-it information to Army lawyers" (*The Army Lawyer*, Aug. 1971, p. 1).

96. See Overseas Basing Commission (note 52).

97. J. S. Robbins, 'U.S. Central Command: Where History Is Made', in D. S. Reveron (ed.), *America's Viceroys: The Military and U.S. Foreign Policy* (New York: Palgrave Macmillan 2004) p. 180.

98. Overseas Basing Commission (note 52). In the case of both Afghanistan and Iraq, all local jurisdiction was removed when both host governments fell subsequent to the US-led invasions. See the comments of M. Cherif Bassiouni, quoted in A. Fahim, 'The Perils of Colonial Justice in Iraq', *Asia Times* (6 July 2005), available at <http://www.atimes.com/atimes/Middle_East/GG06Ak02.html>, accessed 1 July 2008.

99. Krepinevich and Work (note 40) p. 33.

100. This coincides with the established trend of the US military keeping its overseas troops based in isolated compounds away from local residents – a policy which Mark Gillem (2007) has shown inhibits any chance of even the most basic fostering of cross-cultural understanding; see M. Gillem, *America Town: Building the Outposts of Empire* (Minneapolis: University of Minnesota Press 2007).

101. Global Security, 'U.S. Serviceman Accused of killing Kyrgyz Sent Home' (3 May 2007), available at <http://www.globalsecurity.org/military/library/news/2007/05/mil-070503-rianovosti02.htm>, accessed 1 July 2008. As M. Cherif Bassiouni, explains, with most SOFAs, the US has the "primary jurisdiction" and "obligation" to prosecute, and if it fails to do so, the host country gets the opportunity; see Fahim (note 98). Despite various Kyrgyz protests, no prosecution has been undertaken to date in this case, and Kyrgyz President Kurmanbek Bakiyev has belatedly demanded that US forces in the country be stripped of their diplomatic immunity.

102. Democracy Now, 'UN Human Rights Investigator in Afghanistan Ousted Under U.S. Pressure' (28 April 2005), available at <http://www.democracynow.org/2005/4/28/un_human_rights_investigator_in_afghanistan>, accessed 1 July 2008. See also UN News Centre, 'Afghanistan: UN Expert Denounces Abuses in Illegal Prisons' (24 Aug. 2004), available at <http://www.un.org/apps/news/story.asp?NewsID=11703&Cr=Afghanistan&Cr1=>, accessed 1 July 2008.

103. US Under Secretary of Defense Douglas J. Feith, *Transforming the U.S. Global Defense Posture* (Washington, DC: Center for Strategic and International Studies 3 Dec. 2003), available at <http://www.defense.gov/speeches/speech.aspx?speechid=590>, accessed 5 July 2010.

104. Ibid.

105. Ibid.

106. US Department of Defense, *Strengthening U.S. Global Defense Posture* (Washington, DC: Department of Defense 2004).

107. Ibid., p. 8. For a summation of the report to Congress and a discussion of some its implications, see R. Henry, 'Transforming the U.S. Global Defense Posture', *Naval War College Review* 59/2 (2006) pp. 12–28.

108. US Department of Defense, *Strengthening* (note 106) p. 15. Feith's SOFA agreement plans have been challenged on the grounds that they have not adequately consulted the State Department, US Congress or key US allies. Critics include Michael O'Hanlon at the Brookings Institute and former US Deputy Secretary of Defense, John Hamre, at the Center for Strategic and International Studies; see J. D. Klaus, *U.S. Military Overseas Basing: Background and Oversight Issues for Congress* (Washington, DC: Congressional Research Service Report for Congress 2004) pp. 4–6.

109. US Department of Defense, *National Defense Strategy* (note 3) p. 5.

110. Cockburn (note 60).

111. Agamben, *Homo Sacer* (note 65); G. Agamben, *State of Exception* (Chicago: University of Chicago Press 2005).

112. C. Minca, 'Agamben's Geographies of Modernity', *Political Geography* 26/1 (2007) p. 92. See also C. Minca, 'The Return of the Camp', *Progress in Human Geography* 29/4 (2005) pp. 405–412.

113. Many others have pointed out, of course, the manner in which "exceptional events" are "used to legitimate authority to exceptional practices", as Andrew Neal most recently underlines; see A. Neal, *Exceptionalism and the Politics of Counter-Terrorism: Liberty, Security and the War on Terror* (New York: Routledge 2010) p. 1.

114. Minca, 'Agamben's Geographies' (note 112) pp. 93, 94.

115. US CENTCOM Commander General John P. Abizaid, *Statement before the Senate Armed Services Committee on the Posture of United States Central Command* (Washington, DC: Senate Armed Services Committee 2006).

116. M. Neocleous, *Critique of Security* (Edinburgh: Edinburgh University Press 2008) p. 13. On this, see also Neal (note 113) p. 2, where he asks, "How can the sovereign state make exceptions to liberty in the name of liberty, or exceptions to the law in the name of the law" and argues that "the discourse of liberty is used both to oppose illiberal security practices and to legitimate them".

117. By drawing on the twentieth-century history of government in the US and UK, in particular, Neocleous rigorously exemplifies why we should reject any theorisation of contemporary Western society portrayed in terms of periodic episodes of 'emergency' in otherwise 'normal' times.

118. Kennedy (note 72) p. 170.

119. On this point, see J.-C. Paye, *Global War on Liberty* (Trans. J. H. Membrez. New York: Telos Press 2007) pp. 1–3.

120. M. Dillon and J. Reid, *The Liberal Way of War: Killing to Make Live* (New York: Routledge 2009).

121. Neocleous (note 116).

122. I have outlined elsewhere how US military commanders have, in recent years, consistently utilised a powerful and persuasive neoliberal 'risk-securitization discourse' to legitimise command strategies; see Morrissey, 'Closing the Neoliberal Gap (note 19).

123. Or what Jean-Claude Paye defines as a 'permanent state of exception', arrived at from the perpetual invoking of the 'state of emergency'; see Paye (note 119) pp. 40–48.

124. Kennedy (note 72) p. 165.

125. Ibid., pp. 8, 10, 12

126. Benjamin (note 5) p. 283.

127. Kennedy (note 72) p. 167

128. Ibid., p. 169

129. See D. Grondin, 'The New Frontiers of the National Security State: The US Global Governmentality of Contingency', in M. Doucet and M. de Larrinaga (eds.), *Security and Global Governmentality* (London: Routledge 2010) pp. 79–95.

130. Foucault (note 6); see also M. Dillon, 'Underwriting Security', *Security Dialogue* 39/2–3 (2008) p. 309.

131. U. Beck, *Risk Society: Towards a New Modernity* (London: Sage 1992); M. Dean, *Governmentality: Power and Rule in Modern Society* (London: Sage 1999); P. O'Malley, *Risk, Uncertainty and Government* (London: Glasshouse Press 2004); G. Mythen and S. Walklate (eds.), *Beyond the Risk Society: Critical Reflections on Risk and Human Security* (Maidenhead: Open University Press 2006).

132. Dalby (note 19) p. 47.

133. C. Aradau and R. Van Munster R, 'Governing Terrorism Through Risk: Taking Precautions, (Un)Knowing the Future', *European Journal of International Relations* 13/1 (2007) p. 108. Cf. Y.-K. Heng, *War as Risk Management: Strategy and Conflict in an Age of Globalised Risks* (New York: Routledge 2006).

134. Aradau and Van Munster (note 133) p. 108.

135. Kennedy (note 72) p. 165.

"A Day in the Life": A Tomogram of Global Governmentality in Relation to the "War on Terror" on November 20th, 2003

MIGUEL DE LARRINAGA

School of Political Studies, University of Ottawa, Canada

This article shares its title with a Beatles' song in which John Lennon wove together several newspaper stories from a particular day to create "A Day in the Life". As with the idea behind Lennon's lyrics I would like to provide a tomogram of global governmentality by using a specific day's events to examine transformations in the current world (dis)order. On November 20th, 2003, several events occurred, including the bomb attack on the British Consulate and the HSBC bank headquarters in Istanbul; a presidential visit by George W. Bush to London accompanied by antiwar protests; suicide bombings in Kirkuk and Ramadi; an evacuation of staff from the White House due to a "blip" on a radar screen rather than a plane; anti-FTAA protests and clashes with police in Miami; all these events can be used as a barometer to examine global governmentality and the current predicament of the representation of global order. This analysis is framed within the broader context of a questioning of the "eventness" of the event, agency and territoriality in contemporary world politics, as well as the process and significance of dating in representing global order. In doing so it attempts to highlight the tensions between global politics understood and articulated from a sovereign optic and an understanding of global politics as a site of governmentality and transversal struggles in a world where "9–11" and the "war on terror" provide the fundamental markers of current representations of contemporary global social order.

I read the news today, oh boy

—Lennon/McCartney, "A Day in the Life"

Likewise, neither is "everyday life" a synonym for the local level, for in it global interconnections, local resistances, transterritorial flows, state politics, regional dilemmas, identity formations, and so on are always already present. Everyday life is thus *a transversal* site of contestations rather than a fixed level of analysis. It is transversal because it "cannot be reconciled to a Cartesian interpretation of space." And it is transversal because the conflicts manifested there not only transverse all boundaries; *they are about those boundaries*, their erasure or inscription, and the identity formations to which they give rise.

—David Campbell[1]

September 11th, 2001. What is the significance of this date? Or, more appropriately, what does this date confer significance to? One answer to this question was furnished the day after "9–11" by US Senator from Arkansas Tim Hutchinson on the floor of the US Senate:

We all woke up yesterday and prepared to go about our normal business in a world that looked the same as it did the day before. Today, everything is different. The New York skyline has changed and so has the geopolitical landscape of the world. We stand at the violent birth of a new era in international relations and national security.[2]

Senator Hutchinson's understanding of 9–11 is one where the event is seen in absolute and singular terms. It presupposes a pre-9–11 world that is one of "normal business", a world of the quotidian where the way we apprehend the world does not change from one day to the next emphasising continuity and routine. The day after 9–11, however, is one where "everything is different", where the significance that we confer upon the world has totally changed, a world of the exceptional emphasising discontinuity and the uncommon. What is interesting about Hutchinson's re-presentation of 9–11, is that it does not confront the day of September 11th itself. The world is understood in relation to the markers of pre- and post-9–11 where we can make some sense, we can give meaning to, these two "eras", but we cannot, however, face the event *in itself*. 9–11 is seen here as a chiasmus, a "crossing over and reversal" – an event that severs history into two and binds both space and time in a way that projects this event as being universal.[3] In Hutchinson's account, with September 11th, history is violently reborn and "we stand" as witnesses to this rebirth, a "new era of international relations and national security" which can be visually substantiated by the *particular absence* of the Twin Towers from the New York skyline which stands in for the *universal presence* of a new "geopolitical landscape of the world." Of course, one could say that "we" faced this event more than any other event in history, lived it in real time on our television screens from countless angles and through endless iterations. Yet this is not so much "facing" the event if we understand facing in terms of confronting the event with reflection from a particular vantage point. Rather, as James Der Derian suggests, "For a prolonged moment there was no detached point of observation, only tragic images of destruction and loss, looped in 24/7 cycles, which induced a state of emergency and trauma at all levels of society. It was as if the American political culture experienced a collective Freudian trauma, which could be re-enacted (endlessly on cable) but not understood at the moment of shock."[4] It is precisely this traumatic kernel at the heart of the event that can be seen as enabling the pre- and post-9–11 representation of history and the understanding of 9–11 as a singular "violent" origin for a new era of international relations.

There are inherent dangers in re-presenting "September 11th" in this way. In the first instance, in understanding 9–11 in absolute and singular terms, one abstracts from the social, economic, cultural and political conjuncture from which the event violently erupted. In other words, it short-circuits the horizon of meaning to which the event can be associated with. In the second instance, and intimately associated to the above, it enables virtually anything to be justified in its name. Both of these dangers have manifested themselves in the subsequent months and years since the event. On the one hand, attempts to situate September 11th in a broader context can automatically become justifications for the event. A good example of this can be seen in the way in which comments about terrorism, one year after the event by Canadian Prime Minister Jean Chrétien were received. Chrétien, in the span of a week, managed to radically polarise opinion (with the majority of comments going against him) after a CBC interview where he merely dealt with the perception of the north by the south as "arrogant, self-satisfied, greedy and with no limits"[5] and then in a speech at the UN linking, in a general way, terrorism and the growing global income disparity between rich and poor. These pronouncements drew derision from many quarters, most notably in a *Wall Street Journal* article by Marie-Josee Kravis entitled "Why is Jean Chretien so intent on finding a justification for terrorism."[6] On the other hand, attempts to frame 9–11 as an exceptional event have led precisely to the use of exceptional measures. The most evident example, but certainly not the only one as will be elucidated later, being the status of Guantanamo detainees as "unlawful combatants" rather than "prisoners of war" and, thus, not subject to the Geneva conventions.[7]

In view of these dangers, one should thus be wary of seeing September 11th in the light of an absolute and singular event. One could even ask if anything *has* changed? This, of course, is not to make light of the horror of the event, or to denigrate the loss of thousands of innocent people to their families and friends. Furthermore, in no way is it an acceptance of the act as something that should be apprehended as any other daily occurrence, regardless of the amount of people who die daily at the hands of others, and it should hence be deplored as a catastrophic act of violence against innocent people without political, economic or social remittance. In the post-September 11th "for us or against us" climate, both within and without academe,[8] words such as these unfortunately remain a necessary precondition to any discussion on the matter almost a decade after the event. To say that not much has changed also does not preclude the possibility of a series of catastrophic events beginning with the military action of Afghanistan, the invasion and occupation of Iraq, and the eventuality of a further expansion of the US "War on Terror", originally labelled, but quickly retracted, "Operation Infinite Justice", as well as every regime under the sun using the "terrorist" moniker to legitimate the use of force against a myriad of types of dissent. In this context, to say that not much has changed does not forestall discussion on the multiple equations such as that of "security vs. freedom" used to curtail civil rights, or that, oddly enough, between "security vs. sovereignty" used by US Ambassador to Canada Paul Cellucci in his performative call for a continental security community entailing the harmonisation of immigration and refugee policies.[9] Finally, to say that not much has changed should not obviate the fact that the word "security" has become ubiquitous since September 11th and shows no signs of abating over a decade later. Yet this is precisely the point. If anything, September 11th has served as a catalyst to practices and processes that were already in place. In other words, the traumatic experience of the event, trauma used here in relation to the symbolic order into which the event irrupted, was subsequently symbolised – made "sense" of through the prevalent frame – in real-time over the hours, days, weeks, months and years since, by inserting itself into a pre-existing chain of equivalences around "security", spreading like wildfire through the already established practices and processes of security, re-confirming and re-enforcing the latter, and simultaneously serving as conduits through which new equivalences, new practices, new processes have begun to operate. It is precisely the symbolic order into which this event violently ingressed that is at issue here and the governmental rationalities and technologies of security of the ensuing war on terror that are being problematised in this article.

This type of problematisation is not, in itself, original. Many authors have provided readings of the event that caution against an understanding of September 11th as an absolute and singular event.[10] For example, David Campbell cautions against seeing September 11th as the "day that changed the world" by highlighting that, with regards to security policy, both domestic and international, the reaction to September 11th can be seen as a "return to the past" of Cold War thinking.[11] Along similar lines, James Der Derian problematises September 11th as a singular event by introducing media quotes from the first bombing of the World Trade Center in 1993, that could have, in their content, been just as easily written in September 2001. In doing so, Der Derian explicitly addresses the rationale behind the use of these quotes as a caution against seeing the event "as an exception that bans critical thought and justifies a state of permanent emergency."[12] In this sense,

September 11th can be seen as having been predominantly represented and made sense of through a sovereign optic that binds the spatial and the temporal together from the universal gaze of the global: an optic that enables it precisely to be a marker for a permanent state of emergency, an infinite war seeking infinite justice under the sign of the "war on terror."

In this article, I would like to take another tack in addressing this problématique. Instead of problematising September 11th directly and attempting to understand the post-9–11 conjuncture from that event as a point of departure, I would like to propose an examination of this conjuncture from the standpoint of another date, one which has not garnered much attention beyond any of the people directly affected by the events that transpired on November 20th, 2003. In contrast to "September 11th" and that which comes to mind when this date is uttered, "November 20th, 2003" is not infused with universal meaning. Beyond that simple fact, which, in itself, should make us reflect upon the clear indication of the United States as a hegemonic power (in the sense of making its interests universalisable at the level of the symbolic), it should be noted that November 20th, 2003 is not just "any other day." Or, at least, it was not chosen at random. It is a day that congested the mediasphere with breaking news. This news, however, was not of a singular event, but of a multitude of events that are, however, intimately related. From the bomb attack on the British Consulate and the HSBC bank headquarters in Istanbul; to a presidential visit by George W. Bush to London accompanied by antiwar protests; to suicide bombings in Kirkuk and Ramadi; to an evacuation of staff from the White House due to a "blip" on a radar screen rather than a plane; to anti-FTAA protests and clashes with police in Miami, all these events occurring on November 20th, 2003 can serve as a barometer to practices of global governmentality under the sign of "security," sovereign power deterritorialisations and reterritorialisations, as well as sites of resistance to global ordering post-9–11. In this sense, choosing this date does not mean ignoring 9–11 or asserting that it is insignificant. On the contrary, I argue that by looking at such a conjuncture of events one can better apprehend the impact of the event and the subsequent "war on terror" upon the present global order. In other words, it is precisely through the way in which these events are effectively bound together through certain chains of equivalence that are articulated in their representation in relation to September 11th, that a vantage point can be created to provide an understanding of the governmental rationalities of security that underpin the war on terror and the contemporary global order at multiple sites of its production, with the intent of exploring how war beyond the battlefield has been rendered possible as a form of global governmentality through the "war on terror". However, what is important to underscore is that these chains of equivalence are primarily articulated around sites of resistance to sovereign power since, as Michel Foucault explains, "where there is power, there is resistance."[13] An examination of these events on this particular day therefore enables us to provide an analysis of a "Day in the life," as illustrated in the Beatles' song title; a tomogram of contemporary global order. The metaphor of a tomogram; of an image of a slice or section of a three-dimensional object is, I feel, apposite as this slice of life can help us elucidate some of the complex assemblages of sovereign and bio-power *and* some of the various sites and forms of resistance to it through transversal assemblages. Although the idea of the tomogram as a slice of life may suggest a static vantage point temporally, it enables a dissociation between space and time through an examination of multiple and disparate eventful sites that are, however, associated transversally. In eschewing a sovereign interpretation of history via the apprehension of a singular event that binds space and time together as alluded to above, the tomogram seeks to examine events associated to ongoing processes and practices. In this way, it provides a useful tool to examine governmental rationalities and technologies that are, in the present case, the quotidian processes and practices of security that are intimately associated to the war on terror and the acts of resistance to these. In this sense, although associated to a specific date in time, it is not so much the date itself that is significant, but the daily practices of security and resistance that mark contemporary global order. In other words, what is ultimately sought here is an understanding of the production of subjective dispositions in a post-9–11 world under the sign of the "war on terror"[14] – in a context where the idea of war is envisioned beyond the battlefield.

The article first addresses the issue of governmentality and sovereign power drawing primarily on the work of Foucault. Second, as emphasised in the introduction piece of this special issue and in relation to Foucauldian analytics, the article addresses the question of war as a central marker in the articulation and production of social order within the spatiotemporal conditions of exceptionality of the contemporary, post-9–11, moment. Third, these analytics are then deployed to read the representations of the events on November 20th, 2003, as well as the practices that are deployed in relation to these events, in order reveal the way in which an understanding of September 11th as a singular event contributes to the mobilisation of a governmental assemblage informed by the logic of war *via* exceptionalism to defend a particular way of life. This reading also provides a critique of understanding events as singularities by situating the events of November 20th, 2003 within a particular governmental context. Finally, based upon this analysis, the biopolitical question of the form of life that is defended in relation to the global war on terror is raised.

GLOBAL GOVERNMENTALITY AND SOVEREIGN POWER

From a preliminary interest in Foucault's work in the discipline of International Relations around metatheoretical questions of epistemology and ontology in the late 1980s, the next decade saw the beginnings of a sustained use of analytics around governmentality to examine the issue of global governance. The end of the Cold War had served to move the critical debate from the predominance of (neo)realism in the field of IR to ways to account for "governance without government" beyond the (neo)liberal research agenda in IR theory. Based on the work of Foucault as well as the social theory work on governmentality in the late 1980s and the early 1990s by such authors as Nikolas Rose, Peter Miller and Barry Hindess,[15] there was an explicit attempt to examine the way in which, as one of its early proponents put it, "the state and the system of states has, to an important degree, and from very early modern times, been a function of knowledgeable governmentality as much of sovereign territoriality" and that their "governmentalization . . . has been the principal device by which the technologization of the political (and the democratic) has been disseminated globally in the modern age."[16] In other words, we see a concern with an examination of the role of political rationalities and governmental technologies in the constitution and maintenance of international order and contemporary transformations understood under the sign of "global governance". Moreover, in moving onto the terrain of the "international," the above quote also reveals a further concern that will occupy this literature from the outset and which becomes more salient in attempts to think about world order post-9–11 from a Foucauldian frame: that of the relationship between biopower and sovereign power, governmentality and juridico-political power.[17]

Foucault outlines his concept of governmentality *via* his lectures at the Collège de France in 1978 and 1979[18] as a way of understanding the advent of a new rationality of government in the eighteenth century and its attendant governmental technologies. Under the moniker of "art of government" he accounts for what he sees as becoming the predominant economy of power from the eighteenth century onwards; one that is primarily concerned with the "business of knowing and administering the lives and activities of the persons and things across a territory."[19] In contrast to the circularity of sovereign power whose ends can be found in the submission of its subjects to its own authority, in the "obedience to the law," the power articulated via governmental rationalities and technologies has as its ends those things it manages. Governmentality is thus about, "disposing things" rather than "imposing" obedience, of "employing tactics rather than laws, and even using laws themselves as tactics – to arrange things in such a way that, through a certain number of means, such and such ends may be achieved."[20] The ends to which these means are targeted and applied to achieve the "right disposition of things," in broad terms, concern the vitality of the state *via* its population, "the laws that modulate its wealth, health and longevity, its capacity to wage war and engage in labor and so forth."[21]

What is articulated and disseminated here, through these governmental rationalities and technologies, is a particular form of power. It is within this context that Foucauldian notions of biopolitics and biopower can be addressed. Biopolitics concerns those governmental rationalities and technologies with the health and well-being of the population as its target. In contrast to sovereign power which, according to Foucault, is the power to "take life or let live", biopower is the power to "'make' live and 'let' die."[22] Along with what he termed disciplinary power, biopower is also understood as a normalising power in that it works to regulate certain conduct through the production of certain subjectivities. Unlike disciplinary power, however, which individualises in its production of subjects, biopower can be seen as massifying in that it operates and intervenes at the level of populations as a whole. Having lived life as its target does not necessarily mean that this is a "benign" form of power as opposed to sovereign power. One should not understand this strategy of power as an attempt to provide a space to secure life in general but, rather, to produce and secure specific forms of life that, as Dillon and Reid succinctly argue, are "amenable to its sway."[23] Indeed, one crucial aspect of biopolitics is precisely that it is, as Sergei Prozorov argues, not "life-as-such" that is being nurtured, but "only the forms of life that are in accordance with its specific rationality."[24] Moreover, it is within the context of biopolitics that the extermination of populations can be rationalised in the name of the securing of the life of a particular population. Indeed, as David Campbell argues, by having the health and welfare of the population as its referent, "decisions are made in terms of collective survival, and killing is justified by the necessity of preserving life."[25] In order to understand how governmentality and biopolitics can be deployed to understand the current "war on terror," one needs to further understand how Foucault articulates the relationship between sovereign power and biopower in relation to war beyond the battlefield.

BIOPOLITICS, WAR AND EXCEPTIONALISM

Foucault was ambiguous regarding the relationship between sovereign power and biopower. In some instances, he seems to indicate that biopower displaces sovereign power as the dominant economy of power from the eighteenth century onward[26] while in other instances Foucault highlights the way in which these economies of power live in tension with each other and mutually reinforce each other.[27] There is one instance, however, in which Foucault does bring together these forms of power explicitly and which is particularly apposite to the subject matter of this article in terms of elucidating the spatial and temporal contours of the "war on terror": the question of racism and war.

In *Society Must be Defended*, Foucault addresses the question of how the power to kill could operate within an economy of power that has "life as its object and objective."[28] As aforementioned, when power takes life as its referent, threats are understood in terms of the survival of a particular population. It is in the delineating of this population and the way in which this circumscription inscribes race in the "mechanisms of the state"[29] that sovereign power is deployed in the service of the exigencies of biopolitics. In relation to sovereign power then, racism, for Foucault, introduces "a break into the domain of life that is under power's control: the break between what must live and what must die."[30] It is through race then that sovereign power and its right to kill can be brought to bear in relation to biopolitical rationalities that separate, categorise and create hierarchies in the biological. Moreover, the logic that underpins the way the biological is taken into account by sovereign power in this sense is connected to, and ultimately transforms, the relationship of war that Foucault characterises as "if you want to live, you must destroy your enemies."[31] By being articulated biopolitically, this relationship is intimately related to the health and welfare of the population. As Foucault illustrates, "The death of the other, the death of the bad race, of the inferior race (or the degenerate, or the abnormal) is something that will make life in general healthier: healthier and purer."[32] Within this contex, the threat need not be articulated explicitly in biological terms. It suffices to understand these threats as "either external or internal, to the population and for the population" and the vitality of this population being predicated, not upon victory against political adversaries, but upon "the elimination of the biological threat to and the improvement of the species or race."[33] Instead of understanding war in terms of conflict between self-same Others within an international realm, war can be understood as something internal to the social, as an asymmetric threat to the integrity of the population. When one translates this, as in our contemporary moment under the sign of the war on terror, as threats to "humanity" as a whole, then one can understand the articulation between sovereign power and biopower no longer in international terms but as having the globe as its plane of operation. As Vivienne Jabri illustrates, "If biopower is conceived as a generalized terrain of humanity, its logic is the production of political space that extends well beyond the limits of the state, thereby reframing the sphere of the international in imperial terms."[34]

Foucauldian analytics of the relationship between sovereign power and biopower thus enables a spatial reconfiguration of war by understanding the central referent of war-making as the health and welfare of the population as a global biopolitical stance. Another facet of his work on war goes further in both spatially and temporally reconfiguring the meaning of war. In attempting to continue to move beyond an understanding of power based upon a juridico-political model, one of the central questions that Foucault tries to grapple with in the 1975–1976 lectures at the Collège de France, is if "war can serve as an analyzer of power relations?"[35] In addressing this question, the author attempts to trace the existence of a counter-discourse to the "philosophico-juridical" discourse of war as a limit concept and of the Clausewitzian dictum of war being the continuation of politics by other means. What Foucault identifies in his genealogical investigation is that Clausewitz was actually responding to this counter-discourse, a discourse that paradoxically develops concurrently with the establishment of the State's monopoly on violence and thus the cleansing of the social body of bellicose relations by expelling war to the limit of the State.[36] This discourse, that he labels as "historico-political," posits that *politics* is the continuation of war by other means. In other words, under the great juridico-political edifice of the State lies war, which is at once the condition of possibility of this State in a foundational sense in that "the law was born in burning towns and ravaged fields," and in its performative maintenance in that "war is the motor behind institutions and order."[37] In this sense, instead of understanding war as a limit concept; *destructive* of social and political order, war, as with the way Foucault generally understands power, is *productive* of such order as well as of the formation of particular subjects.[38] This discourse identifies war as running "through the whole of society, continuously and permanently" while order is produced through an array of rationalities and technologies that have characterised and continue to characterise liberal governmental order. In responding to how Foucault understands the "distant roar of battle" as continuously underscoring global liberal governmentality Jabri reflects:

> What the above suggests is the idea of war as a continuity in social and political life. The matrix of war suggests both discursive and institutional practices, technologies that target bodies and populations, enacted in a complex array of locations. The critical moment of this form of analysis is to point out that war is not simply an isolated occurrence taking place as some form of interruption to an existing peaceful order. Rather, this peaceful order is imbricated with the elements of war, present as continuities in social and political life, elements that are deeply rooted and enabling of the actuality of war in its traditional battlefield sense. This implies a continuity of sorts between the disciplinary, the carceral and the violent manifestations of government.[39]

It is primarily within the context of the post-9–11 world in relation to the "war on terror" that these analytics have become fashionable due to both the way the nature of the threat has been portrayed and the way, as alluded to earlier, in which it has been responded to through exceptional measures justified *via* declarations of a permanent state of emergency. However, one can also easily trace an understanding of the war as productive of order and particular subjectivities throughout the postwar period. Whether it be the role of anti-communism during the Cold War structuring the Western subject and its identity through discourses of danger, or the role of the war on drugs in producing dangerous subjects and subjects in danger while providing the US with a distinct moral identity in the post–Cold War world, the usage of "war" in these specific historical contexts reveal the way in which war can be understood as productive of social and political life within and beyond the borders of nation-states.[40] Moreover, although the term "war" has been used rhetorically in relation to various social problems from poverty to cancer, some of these usages have been accompanied by warlike practices such as the use of force. Indeed, Michael Hardt and Antonio Negri have traced how the rhetorical use of war from the "war on poverty" in the 1960s by the Johnson Administration to the "war on drugs" in the 1980s, to the contemporary "war on terror," has increasingly been accompanied by more concrete manifestations of "real" wars against "indefinite, immaterial enemies."[41] It seems that it is those discursive usages of war that have blurred the lines between internal and external threats and have thus been framed in terms of exceptional circumstances and called for exceptional measures that have been most suited to turning rhetoric into reality in the use of warlike practices. In relation to both the war on drugs and that on terror, they note, "These discourses of war mobilize all social forces and suspend or limit normal political exchange . . . [and] involve armed combat and lethal force."[42] In this sense, and as explicitly fleshed out above through the work of Foucault, Hardt and Negri see war as having become *a regime of biopower*, that is, a form of rule aimed not only at controlling the population but producing and reproducing all aspects of social life."[43] Moreover, even in the more conventional facets of the war on terror, those that involve military intervention such as in the cases of Afghanistan and Iraq, one can see how the logic of war permeates spaces that have been traditionally sequestered from war itself. For example, in her examination of the "postwar" prison, Judith Butler examines the support by the US administration of the forcible detention of detainees that were only tangentially involved in the war effort. In doing so Donald Rumsfeld justified such actions on the basis of examples in Afghanistan of prisoners obtaining weapons and fighting guards inside the prisons. In this context, the author suggests, "the war cannot be over, there is a chance of battle in the prisons and there is a warrant for physical restraint, such that the postwar prison becomes the continuing site of war. It would seem that the rules that govern combat are in place, but not the rules that govern the proper treatment of prisoners separated from the war itself."[44]

What a focus on governmentality and biopower affords us here, in relation to understanding "war beyond the battlefield," is a focus on the quotidian; a focus on day-to-day practices, even under the moniker of "exceptional times," of State time and the limit concept of war. In other words, as Vivienne Jabri explains, war can be seen in the present context of the "war on terror" as inscribing "emergency into the daily routines of social and political life."[45] It is here that an examination of a number of events on a single day through the tomogram becomes particularly useful to examine how governmental practices articulate themselves in relation to a complex assemblage of sovereign power and biopower in a situation where, increasingly, war not only becomes the rule rather than the exception but, as a consequence, war becomes *productive* of certain subjectivities and forms of social order simultaneously at local and global levels. Analysing the war on terror through the analytics of governmentality and biopower also gives a different purchase to international politics beyond the inside/outside geopolitical logics of sovereignty and the sovereign state system. It reveals the alternative spatial configurations of a transversal politics that has been an increasingly salient feature of our late modern condition and been catalysed to ubiquity in the post-9–11 world. What becomes central to the following analysis is the way in which the biopolitical understanding of the threats to society are articulated through particular representations related to September 11th and the governmental rationalities and technologies that are informed by these representations and which, in turn, performatively produce the present global order and resistances to it. To illustrate this dynamic, I now turn to a brief tomographic examination of some of the events in the temporal slice of life of November 20th, 2003.

THE WAR TERROR AND THE GLOBAL LIBERAL ORDER ON NOVEMBER 20TH, 2003: A TOMOGRAM OF GLOBAL GOVERNMENTALITY

The Bombing of the British Consulate and the HSBC Bank Headquarters in Istanbul

As with the morning of September 11th, 2001 in New York, the morning of November 20th, 2003 in Istanbul was a picture-perfect, sunny morning that saw the early commute interrupted by two enormous, simultaneous blasts targeting the British consulate and the British HSBC bank building. Interestingly, the press portrayed the event as either "Istanbul's September 11th"[46] for those situated in Istanbul and "Britain's September 11th"[47] in the British press, revealing the alternative geopolitical imaginaries through which the event was made sense of – i.e., either in local terms of the city in which the event took place, or from the global gaze of the political and economic possession of a former empire and of a present governmental assemblage of global liberal economic and political interests. Ironically, "Britain's September 11th" would come up again, this time in terms of the attacks on "British soil" less than two years later on July 7th, 2005. These alternative imaginaries are also telling if one looks at how the Madrid bombings of March 11th, 2004 were widely portrayed in the press and academe as the first Al Qaeda attacks on European soil.[48] Here, Turkey's status as being part of Europe, which has been a central European issue for years through the interminable accession negotiations, is in question yet occluded through the politics of forgetting in relation to the Istanbul blasts. In asserting the status of Madrid as the first attacks in Europe, a line is performatively drawn between what is Europe and what is not Europe, where the question of race is not far under the surface as it participates in delineating the contours of the global social order to be defended. We should not, however, see this solely as a decision taken unilaterally by the "West" in relation to a voiceless "Other" of which "Turkey" is part. Within Turkey itself, beyond the grand geopolitical, strategic utility of Turkey as an ally in the war on terror and the financial incentives that this entails, we find those who make sense of these attacks in terms of their own quotidian life and the biopolitical imaginary of the contemporary world order. As one witness to the British consulate bombing exclaimed, thinking of the carnage he had just experienced, "It's not just politics. . . . They're attacking our way of life."[49]

Relating this event to September 11th also occludes the other violent events and forms of violence that surround the Istanbul bombings. These attacks were not only preceded by another set of massive truck bomb explosions five days earlier on two Jewish synagogues, but also a rash of violence including the bombing of McDonald's restaurants and a couple of HSBC branches, the launching of a hand grenade at the US consulate and two explosions at the British consulate visa section.[50] Moreover, Turkey had suffered, in the previous couple of years, from a major earthquake killing a reportedly 18,000 people and from crippling unemployment, itself followed by a significant surge in crime.[51] Understanding this event in relation to an interpretation of 9–11 as a singular event provides a reading of it that justifies an exceptional reading that fuels the discourse and practices of the war on terror and which occludes the structural violence of the global liberal order. Notwithstanding the chronic, quotidian violence that the Turkish population is subjected to, it is only through such a massive act of violence in the last days of Ramadan that Turkey was brought onto the radar of the war on terror. Moreover, it is after these bombings that new, "draconian security measures,"[52] were deployed to safeguard Western interests.

A Presidential Visit by George W. Bush to London

The day of the Istanbul bombings serendipitously coincided with a presidential visit by George W. Bush to London to consolidate publicly British support for the war on terror. Bush's visit was accompanied by unprecedented security, which then became the norm for presidential state visits post 9–11. The highest level of police alert ever in the city of London[53] was accompanied by 5,000 police officers on duty over 14,000 shifts[54] and 250 US secret service personnel that would enable the creation of a rolling "sterile zone" around the president and his motorcade.[55] Although these unprecedented practices of security give us some indication of how the perception of exceptional circumstances post 9–11 informs security practices in what amounts to the deployment of a mobile sovereign territory, what is more telling, in relation to how the logic of war and its attendant technologies are understood as producers of order, are those US extraordinary security demands that were turned down by British Ministers and Downing Street. In anticipation of, not only a potential terrorist attack, but also the massive protests that were forecasted to accompany Bush's visit, the US had requested diplomatic immunity for American special agents and snipers that would allow "shoot to kill" rules of engagement.[56] Also turned down were "the closure of the Tube network, the use of air force planes and helicopters and the shipping in of battlefield weaponry to use against rioters [including] a 'mini-gun' [that] is fired from a tank and that can kill dozens of people."[57]

As forecasted, there were indeed massive protests that organisers claimed to have reached 200,000 people, while police estimated them at around 100,000.[58] These protests were described by the *BBC* as "good natured and non-violent" and from which resulted only 67 arrests including a person who threw an egg at the Bush motorcade who ended up being "the 67th."[59] However, what is most interesting about these protests for the purposes of the argument developed here is not their relative docility in the face of a massive security operation and a war mentality, but the way in which the Istanbul bombings on the same day provided the frame of reference through which the protests and the protesters were framed. In other words, what was, as alluded to above, quickly transformed into "Britain's September 11th", enabled a de-legitimisation of the anti-Bush and anti-war protests on the streets of London. This is astoundingly illustrated by the *Times* editorial the next day in framing the protests against the backdrop of the Istanbul bombings:

> The notion that the President and the Prime Minister were deliberately exaggerating the threat from terror for their political ends, never a persuasive one, has been rendered implausible. The demonstrations that had been organised against Mr Bush yesterday, far from being an expression of the collective will of the British people, looked irrelevant and naive.[60]

This quote would suggest that the only way to read the event was through the sovereign lens of state security and war, in opposition to the will of the people who were hopelessly naïve and living under a false sense of security and who needed to be defended against their own judgement. The massive security operation and warlike footing were, thus, justified to save the population from itself and, moreover, since the protests were anti-war protests, this bombing was a positive proof that the attacks on Afghanistan and Iraq under the aegis of the "war on terror" were justified and necessary. Here we see how the chains of equivalence that are created under the sign of the war on terror between disparate events enables the blurring of the lines between war as a limit concept and as producer of social order, and how governmental biopolitical rationalities become enmeshed with warfighting technical capabilities with the globe as its field of operation.

Suicide Bombings in Kirkuk and Ramadi

The Istanbul bombings took place less than seven months after George W Bush had announced, on the deck of the aircraft carrier *Abraham Lincoln*, the "end to major combat operations" in Iraq; less than three months after announcing that Iraq was the "central front" in the war on terror; and a mere seventeen days after announcing that the "terrorists" in Iraq were being fought there, in order to "not face them in our own country."[61] The events of November 20th, 2003 thus took place within the context of this gradual representational shift from a conventional understanding of war – where enmity is understood in terms of a self-same Other and where war is understood in the traditional sense of combat operations – to a representation of war where the enemy is an immaterial "terrorism" and war is fought on a myriad of "fronts" to protect the "homeland." Another couple of events that were congesting the mediasphere that day were precisely non-conventional attacks in the form of suicide bombings on US ally targets in the Iraqi cities of Kirkuk and Ramadi. In Kirkuk, a truck bomb killed four civilians and destroyed the offices of the Patriotic Union of Kurdistan party that was led at the time by Jalal Talabani who was then holding the rotating presidency of the US-installed Iraqi Governing Council.[62] In Ramadi, a car bomb destroyed the house of Sheik Amer Ali Suleiman killing two of his relatives. Suleiman, the leader of one of the largest Sunni tribes in Iraq, the Duleim tribe, was also a US ally as a member of the US supervised city council of Ramadi.[63] These attacks were part of the increasing violence in Iraq since the official "end to major combat operations" announced by Bush and that, as we know with hindsight, would become the daily norm in Iraq over the ensuing years. As much as this chronic situation was represented as a "front" in the war on terror and explained in terms of protecting the "homeland," many of these acts of violence can just as easily be understood as being directed towards the symbols of political representation of a governing council, imposed by the territorialisation of an occupying force, that was perceived as illegitimate in the eyes of many Iraqis. Notwithstanding the myriad of sources, targets and intentions of the violence, the US forces response to these "anti-coalition activities" was an offensive to demonstrate the effectiveness of US military firepower. On the same day, in explaining the reasons and objectives behind the offensive, Brigadier General Martin Dempsey stated that they "felt that the enemy had begun to act with a little more impunity than we want him to have," and that US attacks – including the use of heavy artillery, battle tanks, attack helicopters, F-16 fighter bombers and AC-130 gunships – were designed to intimidate the enemy by "planting the seeds of doubt in their minds" about overcoming US power.[64] Simultaneously, within the same physical space as these conventional military interventions are taking place, the "international community," through UN Security Council resolutions on Iraq such as 1483, 1500, 1511 and 1546 and through the decrees of the

Coalition Provisional Authority (CPA) (under which the above mentioned Iraqi Governing Council was formed and served), sees biopolitical proclamations for the health and well-being of the Iraqi people.[65] Here again, war-fighting capabilities become enmeshed with governmental biopolitical rationalities as the sovereign "right to kill" is intertwined with the biopolitical power to "make live." Although conventional war-fighting here is indeed aimed at an "enemy" and deployed to shape its behaviour, within the context of an occupation and an evolving insurgency, it is undoubtedly aimed towards shaping the behaviour of the entirety of the population in order to save it from itself and to foster a particular way of life.

An Evacuation of Staff from the White House due to a "Blip" on a Radar Screen

Ironically, November 20th, 2003 also witnessed the obverse side of the war on terror as the White House was evacuated and F16 NORAD fighters were scrambled due to a "blip" on a radar screen that turned out to be nothing.[66] In the wake of 9–11 and the anthrax scare of 2001, this was one of many evacuations of Federal buildings, including the White House, in the ensuing years. What this event, among these others, exemplifies, is the general uncertainty as to the nature of the threat which characterises the post 9–11 world and to the way in which this uncertainty is met through certain governmental rationalities and technologies that, as argued earlier, inscribe emergency in social life through the declaration of exceptional circumstances and circumscribe that which needs to be defended biopolitically. In this instance, logics and practices of war, strategies of surveillance and popular mobilisation combine as a form of governmentality that is oriented towards the possibility of future threats and operates in terms of a rationality of the contingent. This form of governmentality, as Claudia Aradau and Rens Van Munster suggest, governs through risk through "policies that actively seek to prevent situations from becoming catastrophic at some indefinite point in the future. War is mobilized alongside other technologies of precaution in a governmental dispositif to avoid terrorist irruptions in the future."[67] If the examples above give us a glimpse into the types of reterritorialisations of sovereign power that can be seen as part and parcel of a loose, emerging governmental assemblage to preemptively defend a particular way of life, the final event examined will provide the clearest example of what way of life is to be defended.

Anti-FTAA Protests and Clashes with Police in Miami

Apart from the massive anti-war protests taking place in London on November 20th, 2003, there were other major protests on the other side of the Atlantic, in Miami to be specific, against the Free Trade Area of the

Americas negotiations. Since the anti-WTO protests in Seattle in 1999 that represented the first organised anti-globalisation protest on a massive scale, the question of security at summits had become a central concern of state and corporate organisers. Although there was controversy about the use of force in Seattle as well as in relation to the conduct of the protesters, subsequent summits have seen a gradual increase in the show of force and the sequestering of protesters and, in some cases such as the Genoa summit of July 2001, in the use of force. Moreover, an increasing panoply of non-lethal weapons such as tear gas, pepper spray, concussion grenades, rubber bullets and water canons as well as more high tech fare already deployed or in development such as stun guns, laser pistols, sticky foams and heat and sound weapons have been used to confront protesters.[68] In an indication of the blurring of the lines between war and policing, these weapons have also increasingly become of interest to the US military to be used for crowd control.[69] In Miami, the use of non-lethal weapons was accompanied by a show of overwhelming force with 2,500 police deployed in full riot armor in a city that had been closed off to everyone but the police and the protesters.[70] Other aspects of military deployment also accompanied this militarisation of social space. Ironically, the money for this operation came from an $8.5 million rider attached to a spending bill for the Iraq war[71] and, as Naomi Klein reported, the Miami police "invited reporters to 'embed' with them in armoured vehicles and helicopters. As in Iraq, most reporters embraced their role as pseudo-soldiers with unsettling zeal, suiting up in ridiculous combat helmets and brand-new camouflage flak jackets."[72] In the end, the way the protests were handled, with the use of overwhelming and excessive force, was considered a "success." This increased militarisation and strict control of dissent became a model – the "Miami Model" – for dealing with future protests. Another way in which different components of the governmental assemblage addressed above have been conceptually linked can be found in the comment by Miami mayor Manny Diaz in relation to what he perceived as the overwhelming success of the Miami model: "This should be a model for homeland defence."[73] Moreover, we can also find here the drawing of the line between the way of life to be defended and those who are deemed a threat to that way of life. Those who protest against the existence and spread of the neo-liberal order are identified through chains of equivalence that find resonances in the other stories in the mediasphere that day by relating anti-FTAA protestors and terrorists while simultaneously externalising the threat. Indeed, Miami Police Chief John Timoney repeatedly represented FTAA opponents as "punks" and "outsiders coming in to terrorize and vandalize our city."[74] Naomi Klein's account of these events concludes with a point that echoes the way I have attempted to read this event and the other events of November 20th, 2003. In making the link between the global war on terror and the deployment of the Miami model, she concludes: "The war is coming home."[75] Yet one can say that the war has been simultaneously at home and abroad all along. It is a war that can increasingly be understood as drawing the line between a way of life to be defended, and those that are deemed as a threat to this way of life. The chains of equivalence articulated here between anti–free trade protestors and terrorists provide the clearest evidence that the way of life to be defended here is a *neo-liberal* way of life. The increasingly militarised response to any form of contestation to the present order, reveals the way in which war is becoming the predominant frame through which the maintenance of a neo-liberal social order is envisaged.

GLOBAL NEO-LIBERAL SOCIETY MUST BE DEFENDED

Although the contemporary conjuncture is not unique in terms of its power to produce subjectivities within and beyond state boundaries and we may question the teleological dimensions of Hardt and Negri's biopoliticisation of war addressed earlier, as the above analysis suggests, the war on terror does contain elements that are more amenable than other cases in terms of both the breadth and depth of its ability to produce subjects and social order both locally and globally. As alluded to in the introduction, the way in which 9–11 has been made sense of as a singular, extraordinary, event and, consequently, the way in which the "nature" of the threat in the "war on ter- ror" has been articulated – concealed, networked and lethal – as well as the referent to which the threat is supposedly aimed – the "homeland," "free- dom," "civilization" or "humanity" as whole – all lay the groundwork for the exceptional circumstances that inform an understanding of war that is simul- taneously global and local as well as without end, since the immateriality of the threat and of the referent make it impossible to establish criteria through which to delimit the war both spatially and temporally. As addressed above, from a biopolitical standpoint, references to civilisation or humanity creates a caesura between the civilised and the human and the un-civilised and in- human. It is within this context that we can understand the reappearance of sovereign power in Schmittian terms as the power to decide exceptional circumstances and the suspension of the law, but also in terms of decid- ing upon where this division takes place – i.e., the power of determining what is human and civilised and deserving of life and its fostering, and what is in-human and un-civilised and not only left to die, but actively excised from the population *qua* social body. In other words, here sovereign power articulates itself in relation to the logic and aims of biopolitics. Instead of self-same Others as the sources of enmity and as the referents of war in its conventional frame of inter-state relations, the enemy becomes what has been deemed hostile to what is decided upon and circumscribed as "the population" and its existence in relation to the form of order that permits its reproduction and the logistical rationalities that undergird it. As Julian Reid suggests in relation to the war on terror, "it is deemed necessary to defend the logistical life of society from enemies which threaten to undermine the logistical efficiency of society itself."[76] As the examples above highlight – whether it be the representations and reactions to the Istanbul bombings, the security measures of Bush's visit to London and its representational ties to Istanbul, the more conventional aspects of the war on terror such as the responses of the US military to "anti-coalition activities" in the Iraqi "front" of the war on terror, or the creation of the "Miami model" to face anti-FTAA protesters – within the contemporary moment, war is understood in terms of defending a way of life, and this way of life has been articulated in relation to a global liberal order and, in particular – through the tropes of "freedom," "free markets," "free trade" and "liberty" – as a neo-liberal economic order. Indeed, as Latha Varadarajan, suggests, "questions regarding national security are closely interwoven with constant invocations of the desirability of a particular kind of political economy."[77] In doing so, she cites George W. Bush from the introduction of the 2002 National Security Strategy of the United States:

> The United States will stand beside any nation determined to build a better future by seeking the rewards of liberty for its people. Free trade and free markets have proven their ability to lift whole societies out of poverty – so the United States will work with individual nations, entire regions, and the entire global trading community to build a world that trades in freedom and therefore grows in prosperity.[78]

The line drawn around the "population" to be secured is, therefore, one that is inextricably linked to a faith in free trade and the proclaimed inexorability of the project of globalisation – or through a "neoliberal world vision" rendered as "global integration".[79] It is, within this context, and to use Leerom Medovoi's paraphrase of Foucault, that a *global* society must be defended.[80] In addressing the rationalities that go into the drawing of this line, Medovoi characterises "terrorism" in Foucauldian terms as an abnormality, "the naming of a deviant type against which society must be defended."[81] As some of the examples above suggest, the indefinite and immaterial quality of "terrorism" as a threat referent can thus enable the expansion and mobilisation of this abnormality in relation to a myriad of people, groups, movements that are perceived as being against this liberal way of life. Here, it is the identification of a particular culture raised to the level of a universal category within a common global space that provides the basis for creating the caesura between the normal/civilised/human and the ab-normal/un-civilised/in-human. As Medovoi illustrates:

> The war on terror represents the military targeting of what globalization would consider *cultural abnormality*: beliefs, meanings, and practices of any sort that threaten or resist its . . . vision of incorporation into a global

liberal society. The biopolitical distinction would be between "normal" Islam and the "abnormal" kind that "hates our freedoms" and thus our way of life, or indeed between "normal" culture of any kind (defined as a way of life prepared to be fungible and expedient – so that it can join the marketplace of the global economy) and any kind that apparently wills not to be so.

What the events congesting the mediasphere on November 20th, 2003 reveal are precisely the way in which these lines and distinctions are performatively drawn. Through an analysis of a series of events and interconnections on that particular day, I have been able to reveal the way in which conventional spatial categories and geopolitical assumptions are challenged by the war on terror as well as trace the quotidian governmental practices and the transversal politics that are produced under the sign of war both on and beyond the battlefield.

CONCLUSION

Inspired by a day of tracking news and tracing connections, this article has attempted to take the events of a particular day and the performative connections that were made between them to provide a tomogram of the war on terror through the use of Foucauldian analytics around the concepts of governmentality, sovereign power, biopower and war. What this tomogram of November 20th, 2003 has simultaneously provided, is both an examination of how the portrayal of September 11th as a singular event enables the deployment of a governmental assemblage informed by the logic of war through exceptionalism, as well as a critique of such reading by precisely contextualising the events of the alternative date chosen. What this reading and critique has sought to elucidate is the way in which reading the events of this specific day provides a glimpse into how, under the sign of the global war on terror, the line between war on and beyond the battlefield begins to blur in the production of dangerous subjects and subjects in danger globally.

NOTES

1. D. Campbell, 'Political Prosaics, Transversal Politics', in M. J. Shapiro and Hayward Alker (eds.), *Challenging Boundaries: Global Flows, Territorial Identities* (Minneapolis: University of Minnesota Press 1996) p. 23.

2. T. Hutchinson, 'Floor Statement on Terrorist Act', *Press Release*, U.S. Senate (12 Sep. 2001).

3. This was, of course, not just the view of Hutchinson in the days following September 11th. On the same day, conservative columnist George Will proclaimed September 11th as the "end of America's 'holiday from history.'" G. F. Will, 'The End of Our Holiday from History', *The Washington Post* (12 Sep. 2001) p. A31.

4. J. Der Derian, 'The War of Networks', *Theory & Event* 5/4 (2002), available at <http://direct.press.jhu.edu/ journals/ theory_and_event/v005/5.4derderian.html>, para. 9.

5. 'PM says Western world seen as 'arrogant, greedy'', *CBC News Online* (13 Sep. 2002), available at <http://www.cbc.ca/stories/2002/09/12/chretien_jumbo020912>.

6. M. J. Kravis, 'Why is Jean Chretien so Intent on Finding a Justification for Terrorism', *Wall Street Journal* (26 Sep. 2002).

7. White House, *Special Briefing by A. Fleischer*, 7 Feb. 2002. See 'X-Ray Inmates to Avoid Military Courts', *BBC News* (7 Feb. 2002).

8. Within academe, see the work of the American Council of Trustees and Alumni co-founded by Lynne Cheney and Senator Joe Lieberman in 1995. It has been active in attempting to discredit, intimidate, isolate and marginalise voices which are even moderately critical of US policy since September 11th. See, in particular, J. L. Martin and A. D. Neal, *Defending Civilization: How Our Universities are Failing America and What Can Be Done About It*, A Project of the Defense of Civilization Fund, American Council of Trustees and Alumni, Revised and Expanded (Feb. 2002), available at <http://www.goacta.org/Reports/defciv.pdf>.

9. D. Grondin and M. de Larrinaga, 'Securing Prosperity, or Making Securitization Prosper? The *Security and Prosperity Partnership* as "North American" Biopolitical Governance', *International Journal* 64/3 (Summer 2009) p. 82.

10. The more recent critical geopolitical analysis cautioning us is F. Macdonald, R. Hughes, and K. Dodds, 'Envisioning Geopolitics', in F. Macdonald, R. Hughes, and K. Dodds (eds.), *Observant States: Geopolitics and Visual Culture* (London and New York: I.B. Tauris 2010) p. 5.

11. D. Campbell, 'Time is Broken: The Return of the Past In the Response to September 11', *Theory & Event* 5/4 (2002), available at <http://direct.press.jhu.edu/journals/theory_and_event/v005/5.4campbell.html>.

12. J. Der Derian, 'The War of Networks', *Theory & Event* 5/4 (2002), available at <http://direct.press.jhu.edu/journals/ theory_and_event/v005/5.4derderian.html>.

13. M. Foucault, *The Will to Knowledge: The History of Sexuality*, Vol. 1, trans. by R. Hurley (London: Penguin 1998) p. 95.

14. K. Dodds and A. Ingram (eds.), 'Spaces of Security and Insecurity: Geographies of the War on Terror', in *Spaces of Security and Insecurity: Geographies of the War on Terror* (Aldershot: Ashgate 2009) p. 10.

15. See, for example, N. Rose and P. Miller, "Governing Economic Life", *Economy and Society* 19/1 (1990) pp. 1–31; N. Rose and P. Miller, 'Political Power Beyond the State', *The British Journal of Sociology* 43/2 (1992) pp. 173–205; B. Hindess, 'Liberalism, Socialism and Democracy: Variations on a Governmental Theme', *Economy and Society* 22/3 (1993) pp. 300–313.

16. M. Dillon, 'Sovereignty and Governmentality: From the Problematics of the 'New World Order' to the Ethical Problematic of the World Order', *Alternatives* 20/3 (1995) p. 343.

17. See, inter alia, in relation to a broad panoply of issues: M. Hardt and A. Negri's *Empire* (Cambridge: Harvard University Press 2000); W. Larner and W. Walters (eds.), *Global Governmentality* (New York: Routledge 2004); M. Dean and P. Henman, 'Governing Society Today: Editors' Introduction', *Alternatives: Global, Local, Political* 29/5 (2004) pp. 483–494; J. Edkins, V. Pin-Fat, and M. J. Shapiro (eds.), *Sovereign Lives: Power in Global Politics* (New York: Routledge 2004); V. Pin-Fat and M. Stern, 'The Scripting of Private Jessica Lynch: Biopolitics, Gender, and the 'Feminization' of the U.S. Military', *Alternatives: Global, Local, Political* 30/1 (2005) pp. 25–53; J. Edkins and V. Pin-Fat, 'Through the Wire: Relations of Power and Relations of Violence', *Millennium: Journal of International Studies* 34/1 (2005) pp. 1–24; L. Zanotti, 'Governmentalizing the Post-Cold War International Regime: The UN Debate on Democratization and Good Governance, *Alternatives: Global, Local, Political* 30/4 (2005) pp. 461–487; S. Dalby, 'Political Space: Autonomy, Liberalism, and Empire', *Alternatives: Global, Local, Political* 30/4 (2005) pp. 415–441; D. Campbell, 'The Biopolitics of Security: Oil, Empire, and the Sports Utility Vehicle', *American Quarterly* 57/3 (2005) pp. 943–972; J. Reid, 'War, Liberalism, and Modernity: The Biopolitical Provocations of 'Empire', *Cambridge Review of International Affairs* 17/1 (2004) pp. 63–79; M. Duffield and N. Waddell, 'Securing Humans in a Dangerous World', *International Politics* 43/1 (2006) pp. 1–23; M. Duffield, 'Development, Territories and People: Consolidating the External Sovereign Frontier', *Alternatives: Global, Local, Political* 32/2 (2007) pp. 225–246; M. de Larrinaga and M. G. Doucet, 'Sovereign Power and the Biopoltics of Human Security', *Security Dialogue* 39/5 (2008) pp. 497–517.

18. See M. Foucault, *Security, Territory, Population: Lectures at the Collège de France 1977–78* (New York: Palgrave 2007); and M. Foucault *The Birth of Biopolitics: Lectures at the Collège de France, 1978–79* (New York: Palgrave 2008).

19. N. Rose, P. O'Malley, and M. Valverde, 'Governmentality', *Annual Review of Law and Social Science* 2 (2006) p. 87.

20. M. Foucault, 'Governmentality', in J. D. Faubion, *Essential Works of Foucault*, Vol. 3, *Power* (New York: New Press 2000) p. 210.

21. Rose, O'Malley, and Valverde (note 19) p. 87.

22. M. Foucault, *Society Must Be Defended* (New York: Picador 2003) p. 241.

23. M. Dillon and J. Reid, 'Global Governance, Liberal Peace, and Complex Emergency', *Alternatives: Global, Local, Political* 25/1 (2000) p. 128.

24. S. Prozorov, *Foucault, Freedom and Sovereignty* (Aldershot: Ashgate 2007) p. 7.

25. D. Campbell, 'The Biopolitics of Security: Oil, Empire, and the Sports Utility Vehicle', *American Quarterly* 57/3 (2005) p. 949.

26. For instance, in *The History of Sexuality*, he addresses the emergence of the "power over life" as a situation in which "the old power of death that symbolized sovereign power was now carefully supplanted by the administration of bodies and the calculated management of life." Michel Foucault, *History of Sexuality*, Vol. 1 (New York: Vintage Books 1990) pp. 139–140.

27. For example, in his lecture of governmentality, the author explains that "we need to see things not in terms of the replacement of a society of sovereignty by a disciplinary society and the subsequent replacement of a disciplinary society by a society of government; in reality one has a triangle, sovereignty-discipline-government, which has as its primary target the population and as its essential mechanism the apparatuses of security". M. Foucault, 'Governmentality', in G. Burchell, C. Gordon, and P. Miller (eds.), *The Foucault Effect: Studies in Governmentality* (Chicago: University of Chicago Press 1991) p. 102.

28. Foucault, *Society Must Be Defended* (note 22) p. 254.

29. Ibid.

30. Ibid.

31. Ibid., p. 255.

32. Ibid.

33. Ibid., p. 256.

34. V. Jabri, 'Michel Foucault's Analytics of War: The Social, the International and the Racial', *International Political Sociology* 1/1 (2007) p. 72.

35. M. Foucault, *Society Must Be Defended* (note 22) p. 266.

36. Ibid., pp. 48–49.

37. Ibid., p. 50.

38. E. O. Gilbert and D. Cowen (eds.), 'The Politics of War, Citizenship, Territory', in *War, Territory, Citizenship* (London and New York: Routledge 2007) p. 2.

39. V. Jabri, 'War, Security and the Liberal State', *Security Dialogue* 37/1 (2006) p. 55.

40. For the seminal text examining these contexts and issues, see D. Campbell, *Writing Security: United States Foreign Policy and the Politics of Identity* (Minneapolis: University of Minnesota Press 1992). For a recent take, see D. Grondin, 'The New Frontiers of the National Security State: US Global Governmentality Manages the War Machine for (Any) Contingency' in M. de Larrinaga and M. G. Doucet (eds.), *Security and Global Governmentality* (London: Routledge 2010).

41. M. Hardt and A. Negri, *Multitude: War and Democracy in the Age of Empire* (New York: Penguin 2004) p. 14.

42. Ibid.

43. Ibid., p. 13.

44. J. Butler, *Precarious Life: The Powers of Mourning and Violence* (London: Verso 2004) p. 79.

45. V. Jabri (note 39) pp. 49–50.

46. M. Freely, 'After the Bombs', *The Guardian*, 25 Nov. 2003, p. 6.

47. J. Hyland, 'Britain: Media and Government Use Istanbul Bombings to Intimidate Anti-War Dissent', *World Socialist Web Site*, 27 Nov. 2003.

48. See, for example, L. Thieux, 'European Security and Global Terrorism: The Strategic Aftermath of the Madrid Bombings', *Perspectives: The Central European Review of International Affairs* 22 (2004) pp. 59–74.

49. Freely (note 46).

50. Hyland (note 47).

51. Freely (note 46).

52. Ibid.

53. J. Jun, 'U.K.: High State of Alert On Eve of Bush Visit', *Radio Free Europe/Radio Liberty*, 17 Nov. 2003.

54. J. Wilson and H. Muir, 'Fortress London Braced for Anti-Bush Demos', *The Guardian*, 19 Nov. 2003.

55. M. Bright, '"Shoot-to-kill" ' demand by US', *The Guardian*, 16 Nov. 2003.

56. Ibid.

57. Ibid.

58. 'Thousands Protest Against Bush', *BBC News*, 21 Nov. 2003.

59. Ibid.

60. *The Times*, 26 Nov. 2003.

61. This chronology is developed by Charles Peña in C. V. Peña, 'Iraq: The Wrong War', *Policy Analysis* 502 (15 Dec. 2003) p. 2.

62. A. Ryu, 'Suicide Attack Targets Turkish Party Offices in Kirkuk', *Voice of America*, 20 Nov. 2003.

63. Ibid.

64. 'Bomb Kills Four in Kirkuk', *Associated Press*, 20 Nov. 2003.

65. For an examination of the biopolitical dimensions of the Iraqi intervention and their relationship with sovereign power see M. de Larrinaga and M.G. Doucet, 'Sovereign Power, Governmentality and Intervention: Reading Security Council Resolutions as Acts of Declaration', in M. de Larrinaga and M. G. Doucet (eds.), *Security and Global Governmentality* (London: Routledge 2010).

66. A. Entous, 'White House Evacuated After Radar "Blip"', *Reuters*, 20 Nov. 2003.

67. C. Aradau and R. Van Munster, 'Governing Terrorism Through Risk: Taking Precautions, (Un)Knowing the Future', *European Journal of International Relations* 13/1 (2007) p. 105.

68. J. Godoy, 'Rights Europe: 'Non-Lethal Weapons', Tackle Protests Against Globalization', *Inter Press Service*, 26 Oct. 2007.

69. See J. M. Kenny, C. McPhail, P. Waddington, S. Heal, S. James, D. N. Farrer, J. Taylor, D. Odenthal, *Crowd Behaviour, Crowd Control, and the Use of Non-Lethal Weapons*, Human Effects Advisory Panel: Report on Findings, Penn State Institute for Non-Lethal Defense Technologies (2001).

70. R. Solnit, *Storming the Gates of Paradise: Landscapes for Politics* (Berkeley: University of California Press 2007) p. 179

71. Ibid.

72. N. Klein, 'America's Enemy Within', *The Globe and Mail*, 25 Nov. 2003.

73. Ibid.

74. J. Timoney quoted by N. Klein in ibid.

75. Ibid.

76. J. Reid, *The Biopolitics of the War on Terror: Life Struggles, Liberal Modernity and the Defence of Logistical Societies* (Manchester: Manchester University Press 2006) p. 35.

77. L. Varadarajan, 'Constructivism, Identity and Neoliberal (In)Security', *Review of International Studies* 30/3 (2004) p. 320.

78. Ibid, p. 329.

79. L. Bialasiewicz, D. Campbell, S. Elden, S. Graham, A. Jeffrey, and A. J. Williams, 'Performing Security: The Imaginative Geographies of Current U.S. Strategy', *Political Geography* 26 (2007) p. 413; for a critique of this global geopolitics of integration, see also D. Grondin, 'The (Power) Politics of Space: The US Astropolitical Discourse of Global Dominance in the War on Terror', in M. Sheehan and N. Bormann (eds.), *Securing Outer Space* (New York and London: Routledge 2009) p. 113.

80. L. Medovoi, 'Global Society Must be Defended: Biopolitics Without Boundaries', *Social Text* 25/2 (Summer 2007).

81. Ibid., p. 72.

Los Alamos as Laboratory for Domestic Security Measures: Nuclear Age Battlefield Transformations and the Ongoing Permutations of Security

JEFFREY BUSSOLINI

Department of Sociology, City University of New York, Staten Island, New York, USA

Much of the strategic analysis, geographic knowledge/ignorance, and military-security emphasis of post–September 11th America bore a striking resemblance to Los Alamos of the 1980s. After briefly considering the foreign-policy aspects of the work at Los Alamos and some theoretical, historical, and state aspects of the nuclear age that define the domestic space and the population as targets, I focus on the many ways in which the Laboratory has served as a site of development for domestic security measures. In some cases it is the community of Los Alamos at large which seems to be the template for the measures, inasmuch as this is a town adjoining and surrounding the Laboratory. One of the main thrusts of the argument here is that the division between 'foreign' and 'domestic' in contemporary intelligence and military strategy is specious.

One of the most uncanny – well, perhaps I should say canny (as in home-ly, *heimlich*) – aspects of experience after September 11th, 2001, during which time I was in New York City, was not the tremendously unfamiliar characteristic of the time, as if a new era had been entered, as much as the incredible familiarity of many of the discourses in circulation and practices undertaken in the months and years after the attacks. Much of the strategic analysis, geographic knowledge/ignorance, and military-security emphasis of post-September 11th America bore a striking resemblance to Los Alamos of the 1980s that had been unearthed in my prior fieldwork and memories of growing up there. In some cases the same scenarios, causes, and responses were echoed. In other instances measures undertaken or projected at the

Laboratory in the early to mid 1980s became standard government operating procedure. Within the security sphere, notwithstanding the tracking and manipulation of newly emerging and changing conditions, there is something of an 'eternal recurrence of the same' in which old models and practices are repeated, as is demonstrated, for instance, by the US military infrastructure seeking to find a major strategic opponent to fight after September 11th, despite the lack, beyond Afghanistan, of a direct, major state sponsor of Al Qaeda. Within the horizon of the genealogical inquiry here, however, these persistent themes of security are not seen as simple or exact repetitions, but historical dynamics that are also modified in the context of new security pictures, new fears, and new reckonings of world power. This persistence and change is here analyzed in part through Michel Foucault's figure of the *dispositif* or 'dispositive,' which accounts both for the articulation of different elements into a security 'picture' at a given time (such as nuclear arms, space technology, and anti-communism during the Cold War), and for the shifts in security pictures as historical conditions and the interpretation of them change (as today nuclear weapons are still a concern in the overall mapping of security, but in a much different way than during the Cold War). Los Alamos itself demonstrates this type of change, as it has transitioned from secret site to vaunted home of special expert class of nuclear scientists (post–Word War II) to Cold War centre (developing most of the nuclear weapons of the US arsenal) to uncertain institution of the post–Cold War world (nevertheless still clearly tied to nuclear weapons and to security, despite important modifications of the institution and town).

This phenomenon has several salient aspects. First, it shows that although the major strategic doctrine of the Cold War and nuclear showdown with the Soviet Union dominated much of the attention and work in Los Alamos during the 1980s, this did not prevent the active identification of, and adoption of countermeasures against, other strategic threats, including radical Islamic terrorism.[1] In this respect it seems that different discourses, architectures, and priorities of security interpenetrate and coexist with one another in national security institutions. This is little surprise since, despite well-publicised instances of incompetence, these organisations are not one-trick ponies who adhere only to a single rigid paradigm (even given the coercive and distorting effects of national security ideologies within them). Second, this reveals that, although Los Alamos has primarily been associated with the foreign-policy aspects of nuclear armament, as an institution of national security it has also been heavily involved in domestic security responses and policies, as well as more nebulous cultural figurings of security within US life (in other words, being recognised as a place which is *charged* with security as its purpose and bread-and-butter – this applies as much to old films like *The Atomic City*, and its 'gated-town' mystique as it does to the rash of recent news articles about the 'epidemic' of security failings at Los Alamos).[2]

In relation to war beyond the battlefield, this paper uses ethnographic and historical research about Los Alamos to evaluate two aspects: the ways in which the Bomb and the nuclear age rendered the domestic space into a battlefield, and the ways that Los Alamos has been involved in the production of domestic security technologies along with foreign policy ones.[3] As is further elucidated below, Los Alamos is an important site for study of the relations between war and society, between weapons and democracy, since it has had a special position in these relations throughout its modern existence since 1943. It is both tied into world-historical dynamics of the twentieth and twenty-first centuries and is an exemplary site to study a social space accompanied and heavily affected by institutions of security. The full implications of the observations about Los Alamos, and the larger dynamics at work, are ultimately beyond the scope of this paper, but at least looking at the place in terms of how it fits into processes of development of the US security state, especially pertaining to a shift in the standard conception of the battlefield or theatre of operations, is informative.

In order to examine some aspects of the redefinition of domestic and foreign foci in the nuclear age, this paper first considers the foreign-policy aspects of the work at Los Alamos (which upon closer inspection are seen to relate closely to domestic concerns) and some theoretical, historical, and state aspects of the nuclear age that define the domestic space and the population as targets, then pays heed, at least in an initial fashion, to some ways in which the Laboratory (Los Alamos National Laboratory) has served as a site of development for domestic security measures. In some cases it is the community of Los Alamos at large which seems to be the template for the measures, inasmuch as this is a town adjoining and surrounding the Laboratory, so that a number of security concerns unusual for a town of 18,000 have been cast upon it.[4] Thus the argument here claims both that Los Alamos is an important representative point at which to study American security processes and that Los Alamos itself has in some respects occupied a unique position of influence in shaping these processes.

This paper ethnographically examines several aspects of physical security there as domestic security developments later influencing other sites or widespread security practices – in some cases these security measures were pioneered at Los Alamos, in other cases they reflect part of a larger 'state of the art' among national security and intelligence complexes. Los Alamos is further interesting as cultural artifact since, although it has figured and still figures prominently in US national security, it has also been the object of somewhat of a 'field day' of attacks in media and government over the last decades. To a significant extent, the uncertainty and the hostility toward Los Alamos is an epiphenomenon of ongoing malaise and uncertainty about nuclear arms and their place in human society – thus Los Alamos has been totemically trashed due to its inextricable association with them. These factors make it all the more important to try to get a picture of what happens in Los Alamos on the ground – how security has been figured there and how Laboratory scientists and town residents respond to changing conditions and construals of national security. This paper does not aim to 'defend' Los Alamos or to take a polemical position, but, first and foremost, via prolonged and extremely fine-grained ethnographic analysis alongside theoretical interpretation, to give some presentation of what is happening and has been happening there, often in terms of how it relates to other institutions and larger dynamics. Although they are treated separately initially for analytic purposes in this paper, one of the main thrusts of the argument here is that the division between 'foreign' and 'domestic' in contemporary intelligence and military strategy is specious – a kind reviewer of this paper suggested that the fusion term 'intermestic' may more accurately capture this dynamic.

LOS ALAMOS AND FOREIGN POLICY

Los Alamos National Laboratory is well known, even notorious, for its association with nuclear weapons. The place has enjoyed the prestige and the stigma of its involvement with these technologies as they have been a central element in foreign policy and military doctrine. The qualities of nuclear weapons (tremendous destructive power, lingering effects) and the strategic projections of the Cold War situated these as weapons that would be used at an extreme distance, either inter-continentally by bombers, submarine-launched, or land-based missiles, or, at the outside, might be used in Europe as medium- or short-range weapons, as in the much-rehearsed NATO scenario of a Warsaw Pact armored push through the Fulda Gap in Germany – it should be noted that this had the effect of casting European soil as a battlefield, perhaps not surprising in that much of Cold War thinking and doctrine was a direct outgrowth of World War II.[5] Notwithstanding the domestic nuclear industries, testing, and the vulnerability of the domestic population in a nuclear war, nuclear weapons are often presented and discussed as elements of foreign policy almost exclusively. In many respects this focus on nuclear weapons as instruments of foreign policy was crucial to the turning of a blind eye toward the very real, tangible domestic effects of nuclear weapons development.[6]

As a general centre for nuclear science, and collection site for a great deal of specialised expertise, Los Alamos has also participated extensively in international efforts to manage nuclear technology. Thus, a portion of the Laboratory has long worked with the International Atomic Energy Agency on the promotion of nuclear power and on the detection and discouragement of nuclear proliferation, as well as the reliable monitoring of radiological emissions. Weapons science has often gone hand in hand with the development of measuring techniques and practices. As a result Los Alamos technology has been used in Iraq (in the 1990s – as well as in the ill-fated lead up to the 2003 war) and in North Korea to try to locate weapons and isolate a safe nuclear fuel cycle (with no excess weaponised material). Former Los Alamos National Laboratory Director, and plutonium expert, Sig Hecker has visited nuclear sites in North Korea and is a frequent commentator about them. This too is an eminently foreign-policy centred dimension of the work, though similar technologies have been used in monitoring radiation levels at a number of domestic sites as well.

Los Alamos technology and personnel have made up part of the security envelope of both the Atlanta and Salt Lake City Olympic Games, as well as "a number of other similar events, and at a number of government and private institutions," according to Jamie who had worked in the relevant Nuclear Technology group, and later in the Center for Homeland Security (an example of Los Alamos adapting itself to a wider security discourse) – and who travelled to Salt Lake City during the games.[7] Interest in advanced technologies to detect unconventional weapons and dangerous substances, including nuclear material, explosives, and chemicals, was intense at the Salt Lake City games in the years after September 11th. Jamie said that "Los Alamos had been playing such a role quietly but prominently for decades. Given that we were on the cutting edge for scanning and measuring devices for radioactive and other hazardous materials, security along those lines in different critical settings and events has been part of domestic preparedness for decades." Along these lines he said that Los Alamos personnel had worked with a number of different government agencies, and with several different private security firms. He also described how "there's the events or sites that our people were actually conducting the work and measuring, but there's also many other instances where we have cooperated with the building and set-up of systems that will be operated by other institutions or organizations. And beyond that there are counters and devices that we have built and standardized and that are subsequently mass-produced for widespread scientific or security use."

Less widely known but equally important for understanding the history of Los Alamos is the role of the intelligence community in funding and guiding research within some of the divisions of the Laboratory. This history, and the particular bureaucracies around secrecy, are useful in understanding ways in which information and work have been organised in US security institutions.[8] Clearly, information and interpretation on nuclear technology has been valuable to intelligence organisations, and those best-suited to that task have often been the scientists who develop the weapons. As a result, there has always been a strong stake and interest in nuclear intelligence matters at Los Alamos and a corresponding cooperation with other government organisations, especially the Central Intelligence Agency, and, of course, the Department of Energy's Office of Intelligence. Within the Department of Energy 'Q Cleared' section of the Laboratory there is a CIA

'SCIF (Sensitive Compartmented Information Facility,' or additional security cordon that requires the CIA 'SCI (Sensitive Compartmented Information) Clearance' badge). Many Laboratory employees hold a 'Q Clearance' but a much smaller set hold the 'SCI' in addition.[9] It should be noted that, though not unique to Los Alamos, this practice of compartmentalised access has been increasingly applied in the private sector in companies who separate work projects and protect proprietary information.[10]

In addition to employing Los Alamos scientists for their nuclear expertise, though, the CIA has also funded specialised research at the Laboratory for decades to support its mission, largely through promoting the development of technology that facilitates the exercise of 'tradecraft' in the field. Sometimes this technology may be nuclear-related, as in using certain isotope signatures to 'mark' a target that can be traced by it, but it also may be other high-technology devices designed with a specific purpose. The following example is included, although it would seem to bear on espionage carried out in other countries, because it demonstrates how the work at Los Alamos has borne on some objects that appear to be the most mundane and common, and how these technologies come to take a role in later economic developments such as consumer electronics. Nanette, who had worked within the 'International Technologies' Division of Los Alamos, related how an example of such technologies might be:

> Devices that would aid in subtle and unobserved communication in the field. Miniaturizing and disguising certain components to make them seem run of the mill, but retaining a high degree of functionality and reliability. For instance, ladies often carry a make-up compact, that is pretty unobtrusive, and they might well look in the mirror of the small round case and mutter to themselves while applying powder, either about how bad or how good they look! (laughs) We thought, why not put a small and effective transmitter in there, that they could easily use for radio communication. Same might go for a lipstick or for a man with a lip balm or a cigar. This brought us right away to another issue which was range and power. The ideal compact little case isn't so ideal as an antenna if communication needs to be carried out over longer distances or on specialized frequencies. So, we coupled that with a T-shirt or blouse antenna that would have very small metal fibers woven into it, making the overall garment a pretty darn effective long-range antenna.[11]

Although clearly bearing the mark of espionage tradecraft, this example may not seem too outlandish today given the ubiquitousness of compact cell phones, but bear in mind that this was work completed in the eighties, and which likely developed some of the electronics that would later feature in cell phones – which points out one way in which the 'domestic' economic space of the US has been heavily influenced by military and security development. Like the previous two areas, nuclear weapons and international safeguards, this aspect of Los Alamos's work seems definitely intended for foreign policy–type applications in overseas espionage. Hence we can say without too much trouble that Los Alamos is primarily, almost exclusively, known for its involvement with technologies that are seen to be the tools and trade of international relations as undertaken by US diplomatic, military and intelligence organisations – but that these 'foreign policy' aspects are intimately tied to domestic consequences and to everyday life. It is also the case, however, and less frequently noted, that Los Alamos has developed technologies and practices that have found wide application within the US domestic sphere and economy, and which bear on the considerations in this special issue about war away from the formal confines of the battlefield – or about the expansion of the battlefield, if you will, to the extent that 'everyday' practices and situations have become encompassed within it.

COLONIALISM AND POPULATION TARGETING AS DOMESTIC ASPECTS OF THE NUCLEAR AGE

This section considers the intractable role of colonialism, both internal and external, to the nuclear age, as well as the transformation of the home front into battlefield by the population targeting of nuclear weapons. This section draws, as much as space allows, on the analysis of Michel Foucault since nuclear weapons and colonialism occupy decisive places in his reckonings of security and contemporary politics (in biopower and biopolitics).

It is apropos to point out that the nuclear state has been a deeply colonial one (whether it was in the United States, France, Britain, or the Soviet Union) which has exploited or endangered colonised lands and peoples in the acquisition of nuclear materials, the development of nuclear technology, and nuclear testing.[12] Thus the location of these countries' respective testing sites in New Mexico, Nevada, and the Pacific for the US; Algeria and the South Pacific for France; Australia and Christmas Island for Britain; and Kazakhstan for the Soviet Union is telling. And, the New Mexico mining of Uranium using ill-protected Navajo workers in unsafe conditions, entailing many cancer deaths and environmental effects, is parallel to the Canadian Uranium mining at Great Bear Lake, which decimated the Dene and heavily polluted the lake.[13] As Donald Grinde and Bruce Johansen have summed up this legacy for the United States: "The geographic range of purported radiation poisonings spans half the globe – from the Navajos in the Southwest United States, to Alaskan natives whose lives were endangered when atomic waste products from Nevada were secretly buried near their villages, to residents of the Marshall Islands in the South Pacific, an area in which the United States tested atomic and hydrogen bombs in the atmosphere between

1946 and 1958. As investigations deepened, it appeared that the treatment of Navajos was not the exception but only one example of a deadly pattern of reckless disregard for indigenous life – human and otherwise – in colonized places."[14]

It is noteworthy that the Uranium ore used to produce the fissile materials for the first several American bombs (Trinity test, Hiroshima, Nagasaki, Bikini test bombs) came from the Great Bear Lake Port Radium mine and from the Belgian Congo. So it seems that a first domestic aspect of what historian Michel Foucault termed *dispositif de sécurité* ('dispositive of security'/ 'apparatus of security') surrounding Los Alamos and the nuclear age involves colonisation and the extension of sovereignty, often brutal sovereignty, over colonised populations.[15] Grinde and Johansen have noted that "half the recoverable uranium within the United States lies in New Mexico – and about half of that is on the Navajo Reservation."[16]

Rather than a trend or fad, the interpretation of Foucault is important here for two reasons: first, the *dispositif* that he describes and develops in work from the mid-1970s onwards is an important historical concept which is designed to be able to take account of both genealogical persistence and historical change (and we have succeeded neither in understanding fully the historical effects of the nuclear age nor the ongoing persistence and development of the security 'state' – his analysis, specifically devoted to 'security' and largely overlooked in this respect, is thus an important point of reference here); second, nuclear weapons and colonialism are central to Foucault's thinking in the places where he sets out to describe security and biopower. Although there is no systematic theorisation of the Bomb in his work, it comes up repeatedly as a key political technology or horizon when he is trying to understand security, biopower, and biopolitics. Thus, considering crucial passages of Foucault here both contributes to the historical analysis of 'security' in the nuclear age and serves to ground invocations of Foucault in the wider literature that are often free-wheeling and decontextualised. It would not be inaccurate to say that one cannot understand the concepts of biopower and biopolitics in Foucault without taking into account the key role of nuclear weapons and of colonialism.

In Foucault's presentation of biopower (which according to him follows upon and was structured by mercantilism and colonialism), he defines it in one key respect as the power to make live and let die (*faire vivre ou de rejeter dans la mort*), a mutation of the standard notion of sovereignty as the power to make die and let live.[17] He investigates the techniques and logics of government that are devoted to this end. Explicitly following on and expanding Foucault's concept, Giorgio Agamben developed the corresponding notion of "bare life" (*nuda vita*), which he uses to describe those who are exposed to harm through the imposition of an unusual juridical status and the exercise of state power (or, in some cases, the lack of exercise of state power).[18] Foucault describes the racism of the exercise of this kind of power which bears uncomfortably on the colonial examples at hand:

> You understand, consequently, the importance – I would say the vital importance – of racism in the exercise of such a power: it is the condition under which the right to kill can be exercised. If the power of normalization wants to exercise the old sovereign right to kill, it is necessary that it go through racism. And if, inversely, a power of sovereignty, that is to say a power which has right to life and death, wants to work with the instruments, the mechanisms, the technology of normalization, it is necessary that it too go through racism. Of course, by putting to death I do not intend only direct murder, but also all that may be indirect murder: the fact of exposure to death, the multiplication for certain people of the risk of death or, simply, political death, expulsion, rejection.[19]

There is a whole current of scholarship which bears out the relation of what Foucault is describing to nuclear history, some of which is mentioned in Footnote 12. One of the most powerful is Valerie Kuletz' socio-geographic analysis in which she indicates the placement of sites of nuclear danger – testing, development, mining, etc., – has overwhelmingly been on lands of native peoples around the world.

The fate of the Dene of Great Bear Lake and of the Dine (Navajo) and Lagunas in New Mexico seems to instantiate precisely this dynamic. The criminal neglect in willfully exposing miners to radioactive conditions is an exposure to death, a massive multiplication of risk, that was facilitated through the racism of the Canadian and American business-government military structures. In seeking redress from the federal governments to which they were subject, both groups, the Dene and the Dine-Laguna, faced political death and rejection when federal decision-makers and bureaucrats, including the courts, refused to investigate or comment upon their claims. Although they eventually attained some modicum of legal recognition, both were hampered from extraordinary difficulties about their court standing, unreasonable standards of evidence (for instance the challenge that it couldn't be *proven* that the radiation exposure at the mine was responsible for the cancer of any given worker), and a general exclusion that was described by Foucault: the haughty, callous exercise of state power. Thus the experience of exposure to death was compounded by a colonial 'political death' in attempting to engage the federal governments in the problem.[20] A similar point has been made by Maclellan and Chesneaux in regard to French nuclear testing in the Pacific:

> France held to a policy of haughty indifference toward the countries of the region, and Moruroa remained for a long time an absolute priority. There were few voices raised against the policy in France, and most of

those ignored questions of morality and looked at the obstacles testing placed in the way of good relations between France and the countries of the region. French officials seemed happy to ignore hostility to French nuclear tests, expressed every year in resolutions of the South Pacific Forum.[21]

In addition to being deeply intertwined with projects of internal colonisation, the development of nuclear weaponry has also been troubling in the respect that these devices are targeted directly against the population at large, which means that they were always a domestic security issue as much as a foreign policy one – much of the mass psychological effect of the Cold War was along exactly these lines of trying to come to terms with the scale of domestic risk. The nuclear structure of each state held its own population hostage as the stakes offered up to the other side. While the targeting of enemy civilian populations was in no way unknown in the history of warfare, and indeed World War II had seen the massive destruction of civilian populations from death camps, fire bombings, and atomic bombings – nuclear doctrine closed the circle and made populations the pre-eminent target. While Cold War planners in the United States and the Soviet Union of course devoted attention to planning the destruction of the military infrastructure of the other side,[22] the real jugular and backbone of deterrence was the annihilation of enemy populations.[23] Even though this continued earlier trends, it was a unique step. It is precisely this turn that Foucault mentions explicitly near the end of *Il faut défendre la société/ Society Must Be Defended*, when he considers the atomic bomb in relation to bio-power, and notes that the bomb has the power to "tuer la vie elle-même" or to "supprimer la vie," that is "to kill (human) life itself," or "eliminate life."[24] Population itself becomes the vital target. The bizarreness of this situation has been pointed out by Jackie Orr in her *Panic Diaries* where she notes that, in the Cold War, military strength was indicated by bombing oneself, as in nuclear testing: we were bombing ourselves repeatedly (many times over) to prove strength.[25] In the triple-respect of internal colonialism, mutual targeting of enemy populations, and domestic nuclear testing, the domestic space itself became coterminous with the battlefield in the nuclear age.

Foucault considers this to be an aspect of what he calls *racisme de la guerre* or 'war racism.' As he describes this phenomenon it is:

> racism of war, new at the end of the 19th century, and which was, I believe, necessitated by the fact that bio-power, when it wanted to make war, how could it combine both the will to destroy the enemy and the risk that it would take of killing those very people whose life it must, by definition, protect, manage, and multiply?[26]

How to simultaneously threaten an enemy with total destruction of the population, while at the same time protecting and developing the vitality and integrity of one's own population? That is the same question that seems to drive the psychotic General Ripper in Stanley Kubrick's *Dr. Strangelove, Or How I Learned to Stop Worrying and Love the Bomb*, who both launches a surprise nuclear attack on the Soviet Union and is obsessed by the 'Purity of Essence' of the American population. In any event, Foucault is onto something in this formulation. How to understand the role and operation of security when the primary goal of warfare, made eminently possible through technological development, is totally to destroy the enemy population, and thus one's own population? How can a state be said to defend or secure a population by exposing it to the ultimate precariousness? In a bit of modern history which may be unsettling for us, Foucault identifies this mass-endangerment of the population as one of the key aspects of the Nazi state:

> It is not simply the destruction of other races which is the objective of the Nazi regime. The destruction of other races is one of the faces of the project, the other face being to expose one's own people to an absolute and universal danger of death. The risk of death, exposure to total destruction, is one of the principles inscribed among the necessary fundamentals of Nazi obedience, and among the essential objectives of its politics. It is necessary to arrive at a point such that the entire population is exposed to death. Only this universal exposure of the whole population to death could effectively constitute it as superior race and definitively regenerate it in view of races which were totally exterminated or were definitively enslaved.[27]

While Foucault's object of analysis here is the Nazi state, the truly disturbing aspect about this passage is that he sees this as exercising a lasting transformation politically. Here the notion of the universal danger of death is the touchstone of a new type of politics, a new conception of security, however much this may call upon more ancient forms of hatred and barbarism.

The Nazi state and the atomic bomb both figure centrally in Foucault's descriptions of biopower, and as a crucial prelude to the concept of biopolitics that he develops especially in *Naissance de la biopolitique*,[28] where neoliberalism is described as a biopolitics that presented itself as an answer to or critique of the totalitarian states of the twentieth century (New Deal America, Nazi Germany, and the Stalinist Soviet Union). He maintains that this biopolitics, extending from the general population threat and exposure to danger (he says that like prior liberalism this biopolitics may take as its motto to "live dangerously" (*vivre dangereusement*), and identifies the exposure of the population to continual and extreme risk as part of its *modus operandi*), seeks more fine-grained information about the population and more effective modulations of productivity, health, and other population flows.

Foucault is quite clear, though, that the bomb plays a key role in this development by enabling the potential death of the entire population. The bomb – the advent of the possibility of nuclear destruction – along with the genocidal thinking of the Nazi state – figures repeatedly in key passages where he is seeking to outline this dynamic: as indicated, it is a crucial turn in the final parts of *Il faut défendre la société*, the bomb is also a signet consideration in the final chapter of *L'Histoire de la sexualité I: La volonté de savoir* entitled "Droit de mort et pouvoir sur la vie/Right of death and power over life" where Foucault seeks to chart modern transformations in sovereignty in light of biopolitics (in the same book where he also extensively elucidates the *dispositif* methodologically), the bomb is linked to fear of the state in *Naissance de la biopolitique* (where he charts the turn, in the Ordoliberals for instance, that seeks to have found sovereignty in the economy as a means of avoiding the excesses of violent political sovereignty), the bomb haunts *Sécurité, territoire, population* as a vexing problem of modern security and *raison d'État*,[29] and Foucault frequently mentions the bomb and scientists such as Oppenheimer in his interviews and political musings.

NUCLEAR-RELATED STATE TRANSFORMATIONS

This section briefly takes up changes in the state structure that accompanied the nuclear age – especially those transformations that have permanently increased the military and security footing of the US government. The advent of nuclear weapons and the growth of the bureaucracy responsible for them within the country resulted in significant changes in the state structure – namely the growth and modulation of military functions to touch upon many aspects of domestic life and economy.[30] We may recall briefly here the ethnographic vignette from the section on foreign policy in which we saw that secret intelligence technologies, such as clandestine transmitters in makeup cases, paved the way for later commercial innovations in electronics. Clinton Rossiter, Giorgio Agamben, Joseph Masco and Gary Wills all assign a special position to the bomb in terms of the modification of state emergency powers – namely making them more sustained or permanent.

Due to the tremendous destructive power of nuclear weapons, Congress sought to establish safeguards over undue concentration of political power in military hands by creating the civilian Atomic Energy Commission in the Atomic Energy and Espionage Act of 1947. The AEC later became the Department of Energy and both have been responsible for the design, production, maintenance, and ultimate life cycle of US nuclear weapons. While according to democratic theory and historical object lessons it was certainly laudable to seek to limit singular control of these weapons by the military, the establishment of a specialised civilian nuclear weapons bureaucracy operating at sites throughout the nation and in cooperation with many universities had the unintended consequence of militarising many aspects of domestic life.

Los Alamos is a case in point. Most workers there are not part of the active duty military (save some who find themselves on assignment there for various reasons at any given time), and Oppenheimer specifically argued against the Manhattan Project plan to commission all of the scientists as Army officers (once he was persuaded himself, according to some accounts).[31] In fact, up until the recent high-profile changeover to operation by Los Alamos National Security, a private consortium, Los Alamos Laboratory employees had since World War II formally been employees of the University of California. Thus throughout the Cold War and after, they were academic employees of the country's major state university system, fostering an atmosphere and an ideology of university life. However, these employees were largely charged with the development of weapons and working at a facility wholly owned by the federal government. Take with this the close collaborations and exchanges between Los Alamos and some members of science and engineering departments throughout the University of California (UC) campuses (as evidenced by the frequent exchange of scientists, for instance the visiting of UC scientists – or indeed those from other universities – during summer or winter breaks, and the purpose-built University of California San Diego and University of California Santa Barbara UCSD-UCSB building which faces the Administration Building at LANL), and the operation of two other National Laboratories, Lawrence Livermore and Lawrence Berkeley, by the UC, and one can see that in important respects the university system became to some extent a research arm for the military.[32] Although California enjoyed the explicit tie with the nuclear weapons complex, many universities (University of Chicago, Cornell) and private corporations (Monsanto, Dow) actively collaborated in weapons work throughout the DOE complex, and had since World War II. The reach of the nuclear complex into academic and business life is deep, and this is not even to take into account other forms of collaboration and funding by the military branches, the Department of Defense, and other national security institutions.[33]

This phenomenon has three important and related dimensions to it. First, the explicit growth of military and national security institutions by number, size, personnel strength, and funding following World War II, so that a much greater proportion of the government is involved in such concerns.[34] The fact that many of these institutions were tasked with operating covertly often makes democratic accountability difficult. Second, the blurring of explicit lines of the military and of national security institutions through collaboration with universities and private corporations, such that millions of employees of those organisations were in fact working for the federal government on security projects.[35] Masco describes this as a vast military economy. Third, this increase in military/national security work and its diffusion throughout the nation and society tends toward the type of total mobilisation described by the likes of Ernst Junger and Carl Schmitt, in which significant portions of the population are engaged in work pertaining to the military or security industry of the nation. This seemed to be precisely the worry of political theorists such as Clinton Rossiter and Giorgio Agamben.

LOS ALAMOS AS SITE OF DOMESTIC SECURITY DEVELOPMENTS

This fourth subsection takes up specific examples of technologies for domestic security developed or used at Los Alamos, as described by actors involved in their deployment or operation. Looking at them enables a consideration of some aspects of the control and surveillance of domestic space. Physical security has been a pre-eminent consideration around Los Alamos since the beginning of its modern guise in 1943 (and the year before that in its selection).[36] Part of the rationale for centralising the design and assembly components of the Manhattan Project there was the physical remoteness of the site, far from any seacoasts and in terrain that made approaching it difficult. It was desirable that it be a place as far as possible from other towns, especially large urban areas where it would be easier for spies or foreign agents to operate unnoticed. The remoteness and the lack of a great number of nearby neighbours was thought to cut down on the number of prying questions asked.[37]

One aspect of the physical site is the altitude of Los Alamos at 7,000 feet above sea level. It sits on a series of finger mesas that extend from the Jemez mountains into the Española valley, on the other side of the large cañon created by the Rio Grande. Being on the mesas rising several thousand feet from the valley below afforded good vantage on the main approaches to the town, and the mountains on the other side provided a natural barrier of advance from that direction. While some of the peaks there could theoretically have been used by interested parties for an excellent view on the town site of Los Alamos and Laboratory facilities, the number of inhabitants was sparse enough that the mountains were a measure of natural security (besides: the main threats of spying on the project turned out to be internal – aspects of personnel security – which was an important lesson of security in and of itself).

The town and research sites at Los Alamos (which were heavily intertwined during the Manhattan Project, and have remained so but to a lesser degree since) were set within forests that served as a further shroud to the activity going on there (especially in a time prior to satellites). This reportedly made the site more attractive than one in the open with little tree cover (like some of the other potential sites). The combination of distance and forest cover made it impossible to see what was happening up on the mesa from a number of points in the Española valley and from the Sangre de Cristo mountains extending north from Santa Fe. The steep roads, the river, and the trees also made it certain that no large, fast enemy force could approach the area – the terrain itself would slow them down and disperse them.

"The Fence"

In addition to the natural topographical aspects of Los Alamos's area that contributed to physical security, the Army quickly established a number of built measures to enhance security at the site. First and foremost among these was "the Fence," which encapsulated the entire Laboratory, town, and surrounding area. Talk of the "Fence" remains a central aspect of Los Alamos vocabulary today, though it now only encompasses the sites of the Laboratory itself (in terms of what is "inside the Fence" versus what is "outside the Fence"). In its original incarnation, this was a barbed wire fence which surrounded the entire town, with a few heavily-guarded "Gates," the "Main Gate" or "East Gate" at the top of the mesa immediately where the main road (since referred to as the "Main Hill Road") completes its steep climb and comes up onto the mesa, and the "Back Gate" or "West Gate" near what is now Bandelier National Monument, at the point where State Road 4 begins its steep climb up into the Jemez mountains. These gates were heavily guarded during the Manhattan Project (and indeed up until 1957) and the entire perimeter of the fence was patrolled by mounted Army guards. Within this general cordon, further physical and personnel security controlled access to a number of Laboratory sites, including the Main Site in what is now downtown Los Alamos (near Ashley Pond, which is surrounded by the Historical Museum, the (new) County Building, the Community Building, and the Los Alamos Inn).

While "the Fence came down" in 1957, meaning that it was pulled back from encircling the town itself to surrounding most areas of the Laboratory (the Laboratory accounts for some 43 square miles of the 108 total square miles of Los Alamos County), this was more of a remodulation of the physical security than a suspension of it altogether. In some ways the physical fence has been continued with a "virtual fence" of surveillance and personnel security that serves some of the same functions and encompasses the community at large. Decision makers at the time the fence was moved figured that the main emphasis of physical security should be on the "business" part of the town, including Laboratory sites where sensitive information or dangerous material was present. This was apparently motivated by two enmeshing lines of thought: that it would be more "economical" to focus physical security on the most necessary areas, rather than continuing it around the town itself, and, second, that the physical town site of Los Alamos tended to be quite secure already, by virtue of its physical topography and the types of people inhabiting it (who tended to self-select for types of work involved in the town, and who would likely look for and report suspicious activities).[38]

According to Don, a Los Alamos "Old Timer" who had been involved in physical security,

> We really didn't need the "Fence" anymore, not as it was during the War around the whole town, anyway. That lasted until 1957 because I think that aura of the War still held over the place, somehow it just seemed to folks that it should be the secret place, walled off and all that. But, since '45 everyone had knew about the place. We realized that, from the standpoint of protecting secrets and materials, it would be a better idea to draw the fence back, it would be more economical to focus it on the Laboratory sites rather than keep the damn thing about the whole town, and splitting the guard staff so that they'd spend so much of their time patrolling the "town side" of the fence, where the biggest problems they'd come up against was kids out drinking in the cañons. It was one thing when the whole place was secret and the Army was handling and footing the bill for the security – they might as well have all those guys out there on horseback! But, as we were redesigning the security situation, we realized that it was far preferable to have those guards we had focussing on the goods – not looking for Johnny and Jane vomiting over in Rendija Cañon! By decreasing the fence size, we could increase the amount of attention paid to each part of that fence.
>
> On top of all that, we damn well realized that the town itself didn't really need that level of security over and above other measures. For one thing, there was a police force in the town that handled all the normal order and law issues. And the size, composition, and training of the police force was quite sophisticated for a town this size. And, of course you could justify those police for the Lab, but shit for the most part they were only dealing with jaywalking, speeding, and maybe noise complaints! The police could perfectly well deal with the town while the Lab concentrated its security on its own sites.[39]

Don's recollections shed some light on a key turning point in physical security in Los Alamos, and also illustrates one of the perennial considerations of reckonings about security, whether it be physical, classification, computer, personnel or otherwise – the balance between depth and breadth in security measures. While the Manhattan Project security reasoning held that it was best to physically isolate and patrol the entire secret facility, later into the Cold War security planners realised that it was simply a waste of resources to protect an entire physical area where no classified or sensitive work was taking place. Breadth of security always trades off against depth, given the same amount of resources at hand.

Even during the Manhattan Project, when Los Alamos was described as "the most secret and secure facility of all time" there were notorious incursions that sprang from the inability constantly to patrol the entire fence. Physicist Richard Feynman drew the ire of security personnel by repeatedly exiting through holes in the fence then turning back up at the Main Gate for re-entry, even though no 'departure' had been logged for him.[40] Such pranks and concerns, which served to demonstrate the impossibility of protecting such an extended physical frontier, became even more common after the war, when the town grew and there were more youth who flaunted the rules. As Don explained, this made little sense from a security point of view, and the vital secrets of dangerous materials of the Laboratory were better-protected by a focusing of security measures.[41]

Significantly, Don also explained that the pulling back of the fence in Los Alamos became somewhat of a model for other protected federal facilities, including other laboratories and military bases. The same reasoning was applied to focus on tighter and more-guarded zones instead of vaster ones containing little or no sensitive work. Don said that "in some ways this is a rule as old as time: you focus on the goods, the really important stuff. But I suppose we had to learn that again, and that became something that was adopted at other sensitive facilities – restrict the size of the protected area and concentrate your resources there. Don't use manpower and money patrolling space that is going to get you little actual protection." He also noted that there is a parallel here with classification, where the restriction of too many documents can interfere with the protection of the ones that are truly most valuable. He said that, even though one might think it is always best to close off more physical area or to classify more documents, in actual security practice it is preferable to strike a balance which focuses on truly important sites and information.[42]

Even the "Fence" around the forty-three square miles of Laboratory ground contains more forest, sage brush, and open area than it does vital sites, but it establishes a first cordon of security and of exclusion, within which individual sites are more heavily protected by fences or other measures. This principle of "concentric security" is hardly new to Los Alamos, and was used, for instance, in the design and defence of ancient and medieval cities, with nested levels of security and protection down to the central and most-guarded spaces of the city, usually the ruler's stronghold. What was characteristic about the way this doctrine was applied in Los Alamos was its alignment to contemporary conditions and its integration of other forms of security (sensors, personnel security, computer security) to provide for a graded security landscape accommodating different levels of security and different types of work within it.

The forty-three square mile cordon of "Fence" in Los Alamos has itself given rise to its own misconceptions about security there – including some of the rash of recent stories about "loose" security. Even though the entire area is technically off-limits and is marked by "No Trespassing: U.S. Federal Government, Department of Energy" signs, the fact of the matter is that most of this area contains nothing sensitive and is basically indistinguishable from similar forest and brush areas right across the fence. Matt, who grew up in town and lived there some in the years following high school, said that the "[federal] government land was my personal playground. The cops couldn't go on there, so a lot of nights you would find us out there drinking, shooting the insulators off the electric wires with a .22, or smoking weed. Of course we weren't over by any of the sensitive facilities, but there's plenty of land out there with no one around."[43] But, the presence of the fence, together with the historical reputation of Los Alamos, has encouraged the belief that crossing this outer fence is tantamount to gaining access to a nuclear weapon or to an inner sanctum of the US security infrastructure. Several would-be adventurers have thus crossed this fence in daredevil acts of exploration. One of these was a reporter for *Wired* magazine who crossed the outer fence, and let on that he had accessed the very heart of the weapons complex. This was described by Roger, a physical security specialist at Los Alamos:

> Yep, a few years back we had this jackass Noah Shachtman sneaking in here, crossing the outer boundary fence and then writing up a story like he had made off with a W-88 or something. I would be the first to admit that we have got problems and challenges with the security here – it is a formidable job. But, what this guy did was hardly a major threat to national security. He only crossed the outer fence in an area far from any active Lab sites, much less any sensitive sites. What he did was tantamount to walking down Pennsylvania Avenue in Washington and claiming that you had broken into the White House. You can be damn sure that if he tried to continue that dangerous little stunt at the Plutonium facility or another sensitive site, he would've been either shot down or arrested, and either way it would have been sad, preventable, and a big mess for his family, his paper, and the Lab.

> He walked into a part of the Lab called (Technical Area) 33, which had been a very secure area, but some decades ago. The whole area had been reconfigured and the sensitive projects there had been moved elsewhere, largely because of concerns about proper protection and security. What he apparently didn't realize was that the very same area he entered – if he even entered it, you'll notice that all of the photos are from *outside* the fence – had played host to girl scout troops in the years before his escapade. You can be relatively sure that we were not going to let girl scouts or brownies into an area that contained either hazardous material or sensitive secrets![44]

This tale illustrates a very interesting aspect surrounding the discourse about security in Los Alamos – the difficulty of separating fact from fiction, the mundane conditions from the mystique. Many security experts would point out that the mystique can itself constitute a powerful aspect of security. As is abundantly clear, Shachtman was seduced by the mystique (and the bandwagon of stories portraying Los Alamos as lax). While his observations are not entirely unfounded, and indeed the area that he accessed had played a decisive role in the development of depleted-uranium munitions and other experimental weaponry during the Cold War, the place was quite tame by the time he visited (notwithstanding the possible lingering radioactive and chemical contamination that afflicts much Lab property). In fact, another security specialist at the Laboratory pointed out that "the area in question, 33, had been so thoroughly reconfigured that there hadn't been any guard personnel posted there since the late 1990s. Thus it would appear that his claim to have sneaked in while the guards were less than 100 yards away was simply a lie – unless one of the infrequent drive-throughs happened to be taking place while he was there. A few decades before, that was one of the most heavily-guarded parts of the Lab, by 2003, not so."[45]

So, the "Fence" around Los Alamos has in no small part been associated metonymically with the security and integrity of the place overall. While the movement of the fence back from the entire town to just the Laboratory land was a significant step, and one motivated itself by security calculations, even that reduction of area under protection left a vast area that would be difficult to protect without a massive dedicated security staff, and likely a military one. The outside fence provides one layer of physical security that is buttressed by others. The most sensitive sites at the Lab include additional fences as well as the guards ("to knock out") and sensors ("to deactivate") to which Shachtman makes reference. The point here is not that the security at Los Alamos is infallible – indeed security folks will tell you that no security is perfect, just more difficult to defeat, and that the more area, secrets, and bureaucracy you have the more difficult the job is – but that the organisation of security there and the reputation of the place can easily lead to misinterpretations about the relative degree of surety there. In fact, former Los Alamos guards told me of several instances in the 1970s and 1980s when would-be trespassers to TA-33 (the same now disused area that Shachtman accessed) were apprehended (in some cases almost shot). This seems to indicate that, within the concentric security model, this was a site that had been subjected to greater or lesser degrees of patrolling and securitisation varying with the sensitivity and danger of the projects carried out there. In fact, while the site had formally been protected by hi-tech sensors as well, these had been moved as the site was reconfigured.

Within the general outer "Fence," a number of sites of the Laboratory are surrounded by additional fences or physical barriers (as well as additional surveillance and guarding) to make unauthorised entry more difficult. Sites containing sensitive nuclear material like plutonium, or secret information like weapons designs or intelligence material are subject to especially intensive security measures.[46] Two of the sites that have involved some of the heaviest security at Los Alamos are the Plutonium Facility (TA-55) and the Criticality Facility (TA-18). Both of them provide interesting case studies into the security infrastructure there and the ways in which Los Alamos's security has served as a model for other sites. Both of these sites are adjacent to Pajarito Road, which passes down through Pajarito Cañon and is one of the main thoroughfares linking the main part of Los Alamos town site to the smaller residential area of White Rock. While this road is on Laboratory, federal government land, it was long open as a public roadway – which could be closed, however, at the discretion of Laboratory security (indeed it was closed, first intermittently then permanently, after September 11th, 2001). The fact that this was a public road yet contained some of the most sensitive facilities at Los Alamos presented clear security challenges and required the adoption of a number of novel measures over the years. A first aspect to be borne in mind is the fact that, notwithstanding the officially open character of this road, a number of measures of surveillance and community security served to restrict the flow and access along this road. Part of this is along the lines of the "virtual Fence" concept discussed previously, where the drawing back of the physical fence did not mean that in other respects this barrier was not maintained through more "passive" security measures.

TA-55

TA-55, the Plutonium Facility, houses apparatus for the machining and metallurgy of plutonium and other fissile material, storage vaults, and equipment for nuclear chemistry and the production of weapons components. It is here that the construction of plutonium "pits," formally halted since the early 1990s, has commenced again.[47] As the site of weapons components and large amounts of fissile material, this has been one of the most heavily secured facilities at the Laboratory. It is located behind several layers of barbed wire fences and a dedicated proportion of the Lab's guard force is continuously present here. In addition to these measures TA-55 has also been on the cutting edge of the use of sophisticated sensor technology to detect intrusion. Different types of motion or heat detectors were used between the levels of fences in order to take note of anyone – or anything – passing between. One security worker at Los Alamos who had worked extensively at TA-55, Juan, pointed out that "we usually get most of our alarms from when animals would make their way through. But that is the price you pay for effective security. If you go out for 100 calls for a coyote or a rabbit, on the 101st you might find someone trying to sneak through. Anyway, these sensors were designed to work in tandem with the other security to make penetrating the facility as difficult as possible."[48] John, who had worked on TA-55 security as part of his purview while working on physical security in Los Alamos for several decades, pointed out that "several of the sensors and systems deployed at (TA-)55 were home grown, in that scientists from the lab, usually in P (Physics) division, but sometimes from IT or another group, were the ones to develop and field them. Sure there were some ideas, some models that weren't up to snuff, but there were others, major advances in the state of the art, that were developed here, then deployed here and elsewhere."[49]

This TA-55 security worker, John, also pointed out that the sensor technologies developed and put in place there had later been adopted by other government and private high-security facilities. He said that this hadn't been widely publicised, "since that is the nature of the business to keep a lot of developments quiet, close to your chest instead of crowing about it, in order to preserve the edge as much as possible." Nonetheless, he said, among the somewhat restricted community of those working in such security settings, there was "a good deal of exchange of information and personnel, and the TA-55 sensors were also adopted in places like submarine bases, government buildings, and banks." As Los Alamos was responsible for so much sensitive material and secret information, he continued, it made sense that it also had been on the forefront of physical security measures and devices. He had several times taken part in teams that were responsible for seeking out, developing and implementing new security measures to increase surety at the Lab.

TA-55 is also noteworthy for a major physical security modification that served as a direct model for further changes adopted at the White House, federal and private buildings, and at US embassies. In 1984, after the bombing of the US Marines and French Paratrooper barracks in Lebanon in 1983, security officers at Los Alamos decided that TA-55 was vulnerable to a car bomb-type attack or physical assault. As a result, they embarked in a multi-million dollar upgrade in order to move Pajarito Road several hundred metres further away from the structure, and to put physical barriers in place to prevent any motor vehicles – especially a car bomb or truck carrying would-be infiltrators – from being able to gain easy proximity to the Plutonium Facility. Some of these physical barriers took the now familiar form of concrete emplacements and movable heavy steel roadblocks.[50] Don explained that "given the threat picture that we were working with at the time, of possible attack by terrorists, it was absolutely essential to make access or proximity to the Plutonium Facility more difficult. We understood that car or truck bombs could pose a major threat, and that if such a device were detonated and breached 55, the potential spread of radioactive material could be catastrophic. As a result we realized that it was absolutely essential to move Pajarito (road) and to further enhance the physical security profile of the building and surroundings."[51] Since Pajarito Road was, at that time, still an open public roadway (owned by the federal government but in general left open to public traffic), security officers saw no option but to move the road away from the building – since moving the building itself was judged to be more impractical due to cost, difficulty, protection of nuclear material, and proximity to the edge of the cañon.

Don, the security worker from Los Alamos, explained that many of the measures and calculations adopted at TA-55 were also the model for similar emplacements elsewhere. He said, for instance, that reckoning of the number, strength, and material composition of the barriers was the outgrowth of detailed calculations as to what would guarantee the security of intended targets from an attack (or at least make it as difficult as possible, or deflect some of the force of such an attack). He commented that "looking at those concrete barriers and such, you might think that they were put in place roughly or haphazardly, just to serve as another visible barrier. Although the appearance of increased security that they provide is an important part of the process, the fact is that they were chosen, built, constructed, and placed according to pretty precise plans. After all, we have got a lot of accumulated expertise here about explosions and shaped charges." He also continued to describe how "these same calculations, or variations of them, were also used in establishing the physical security upgrades at the White House, at sensitive federal or private buildings, and in some military establishments. I was part of a interagency team that worked on the White House and other facilities that were putting in place the same kind of measures." Don also commented that "had similar measures been in place in Oklahoma City, a federal building that hadn't received the same kind of upgrade, the damage would have been far less significant than what we saw on that day. Most of the force of the blast would have been absorbed by the barriers or the air between them and the building, and far fewer lives would have been lost."

TA-18

Also known as the "Criticality Facility," TA-18 played a role in the work at Los Alamos from the Manhattan Project until recent years, originally in explosives testing, then in criticality and radiation experiments from 1945 on.[52] This site was chosen for its location on the floor of Pajarito Cañon, in a relatively narrow passage, with high cliff walls on each side. The cañon and cliff walls would provide natural shielding in the event of explosives or criticality accidents. This desirability in terms of natural protection also simultaneously constituted a security difficulty, as the low-lying site with high ground on each side made for a difficult tactical environment in defending the site, especially since it is immediately adjacent to Pajarito Road (though a rock outcropping that the road follows does partly isolate the facility from the roadway). The combination of sensitive work and proximity to the public roadway has meant the TA-18 has also been subject to a series of high-security measures. Although the road wasn't moved there, the site was "hardened" like 55, with the addition of reinforced concrete barriers, steel roadblocks, and countermeasures. One salient feature of security during this time was the establishment of a heavily fortified guard post on the rim of the cañon opposite the site and above Pajarito Road, affording a view and angle of fire on the road and onto the site below. According to Juan, a guard who had worked there, this post contained "machine guns, heavy machine guns, anti-vehicle munitions, and advanced electronic sensors including infrared and motion devices to scan the cañon below." He recalled CIA security specialists visiting the site to take notes on the design and specifications for that emplacement after the 1993 attacks at the front gate in Langley.

The post described by Juan was one measure intended to redress the vulnerability of the TA-18 site. Another important aspect of physical security was the establishment of a rapid response force near TA-18 and TA-55 that could respond within minutes to an attack or crisis at either site. Intended as a layer of physical security beyond that of the normal guard force in Los Alamos, this team was composed of more highly trained individuals. Seth, who was a member of this team, had previously been part of the Army Rangers, Special Forces, and Delta Force. He said that "several members of the rapid response team had a special operations background. There were some former Delta guys, Green Berets, SEALs and ex-Rangers. On top of that there were others who had been Police SWAT guys or part of other government agencies involving sophisticated tactical operations."[53] Although the career trajectory of special operations-trained personnel into private security organisations in the United States is now familiar, the establishment of this force was an early and concentrated effort to take advantage of such expertise in augmenting security at Los Alamos.[54] Seth noted that he "knew of a number of other guys to leave the service and take positions in various defense contractors or security companies, but this was the first instance I knew of where there was a specific effort to draw in and concentrate expertise like that. It would have been difficult, or, shit, impossible to generate such a team from the ground up, having guys with training and experience like that under their belts, especially counter-terrorist and close-quarters training, was absolutely crucial to the mission we have here." This was particularly noteworthy given that the Los Alamos rapid response force was established not long (within 8 years) after the formal establishment of the Delta Force itself by the Army (Seth described Delta Force training and tactics as the centre of the RRF training).

The establishment of the capability for quick response by highly trained paramilitary troops at Los Alamos is akin to the "concentric security" model discussed earlier with regards to the fence. While the general physical perimeter of the Laboratory itself, and of specialised sites within, is patrolled and guarded by the guard force at large (itself a quasi-paramilitary organisation, using weapons like automatic rifles, assault shotguns, armored vehicles, and heavy weapons), the rapid response force was intended as a harder hitting, specialised team that would overmatch would-be infiltrators in terms of training, tactics, and experience. Instead of general patrol and everyday security, they were intended as a response to specific and dire threats (like the military units it was modelled on). The addition and deployment of this team, at a specialised security facility on Pajarito Road, was seen to add a crucial layer of security to the system established at Los Alamos. Seth's colleague David explained that "the particular combination of dangerous material and expertise at Los Alamos make it a highly valuable target for those interested in obtaining nuclear weapons, material, or knowledge. For a long time during the Cold War, the security infrastructure was organized largely against foreign attack or infiltration by military units, as well as against possible threats from organized violent protest. Over time, especially in the late 1970s and early 1980s, it became increasingly evident that Los Alamos would be a prime target for terrorist groups, either in terms of a catastrophic attack on the site or in theft of weapons or material that they would use in other acts of destruction."[55]

As a result of that shift in threat-assessment, physical security at Los Alamos was shifted to include more emphasis on mobile and flexible response to complement the fences, sensors, barriers, and patrols that constitute much of the security apparatus there. One major shortcoming of the new flexible rapid response ability at LANL was demonstrated in exercises at TA-18: a sufficiently stealthy assault would prevent the rapid response force from being called in the first place, and thus defeat that measure by circumventing it. Several times, including in 1997 and 2002, US Army Special Forces teams operating as mock assailants successfully infiltrated TA-18 and defeated the guards and countermeasures present without the alarm being sent out to the RRF which would trigger their deployment. This shortcoming was cited as a major reason why Pajarito Road was permanently closed to public traffic in the years after September 11th, 2001, and the major functions of TA-18 were closed down and moved to other sites. Despite these highly publicised security failings, it should also be noted that other similar exercises resulted in the stopping of the infiltrating force – sometimes by the regularly posted guard force and sometimes by the rapid response force which had been tipped off by sensors or by on-site guards. David noted that "there were certainly times when they (the mock aggressors) got through, but there were also others when we stopped them. In fact there were more times when we stopped them, but, just like you hear, 'they've only got to be successful once, we've got the be successful all the time.' "

CONCLUSION

Los Alamos occupies a peculiar yet representative place within the architecture of US security. Central in some respects and yet peripheral and downright out-of-date in others. Looking at the place, in terms of some peculiarities of its operation as well as in light of some of the larger dynamics heralded by the nuclear age, shows several deep respects in which prior conceptions about the battlefield and the homeland have been reorganised as to overlap considerably. With the advent of the atomic bomb, earlier configurations of the foreign and the domestic were redrawn. The homeland became the exclusive target as well as the site for dangerous production and testing of the weapons. To address then how war beyond the battlefield comes to spatialise the homeland *as* battlefield in studying the case of Los Alamos, one turns to a biopolitical history of the nuclear age that connects the Cold War to the War on Terror through the aegis of homeland security. This shift from national security to homeland security is henceforth achieved through the enactment of domestic security measures deployed at the local scale but aimed at a national and global one. Moreover and as has been argued in the paper, colonialism deeply influenced this process, both with the internal colonisation of native lands and peoples within nation-state boundaries (as Navajo in the Southwest and Paiute and Western Shoshone in the Nevada Test Site, or Kazakhstan for the Soviet Union) and the external colonisation of additional territory in other locations (as Bikini for the US, Algeria and Moruroa/Fangataufa for France, and Christmas Island for the UK). If the nuclear age has dramatically increased the risk at home, it has also especially visited these risks on colonised and vulnerable populations. Colonialism has affected each stage of the process of developing nuclear weapons, from uranium extraction to location of research design centres and testing sites. The scale of nuclear plans and nuclear destructive power also served to place the population at large in danger of death, one of the disturbing parallels to the Nazi state. In this we might recall Hannah Arendt's suggestion that totalitarianism is the 'returning home' to Europe of practices long used 'abroad' in colonies by European nations. In the turn to population targeting,[56] the battlefield becomes coterminous with the homeland, and all residents share a risk of imminent death. Such a situation is bound to produce political changes. Among them are the militarisation of the internal space and economy of the nation, the transformation of the state structure towards secrecy and bureaucracies of security, and the intensification, or permanence, of states of emergency or of exception (as the temporal and conceptual boundaries of being 'at war' or 'at peace' were collapsed to minutes from weeks or months).

Regarding some of these larger issues of the world-historical changes brought by the bomb, Los Alamos is almost peripheral – except for the fact that it is the place that most directly ushered in this state of affairs to which it maintains an ongoing tie. As such the place has a particular relation to these dynamics. It sits on lands partly condemned from Neuvomexicano home-steaders and two Pueblos. And, regarding risk of the home population – did not the workers there take significant risks in *living* there with their families? Despite the scientific rigour and safeguards of the place, working with such materials and processes entailed risk of sudden catastrophic accident or of leakage or exposure of toxins over time. The local governance of Los Alamos is affected by the fact that nearly half the county is federal land and that federal institutions and agencies operate there. The story of the assertion of political independence and representation by the City-County of Los Alamos is itself a fascinating one, which has been detailed by historian and former state representative Marjorie Bell Chambers.[57] It is little wonder, then, that the place exhibits informative examples of the redefinition of domestic space in the nuclear age and successive reckonings of security. It makes sense that security measures or techniques there would find application or echo elsewhere, either as evidence of Los Alamos's influence or of a generalised network of security institutions and practices that have taken root throughout the US (and many other countries and territories). It comes as no surprise that technologies developed at Los Alamos to monitor international weapons research should be used as domestic security measures. It may be less known, but perhaps equally expected, that measures for the securing of physical space, personnel, and computer systems developed and implemented there have been picked up throughout the security infrastructure. Despite being the object of controversy and derision over the last decades, Los Alamos remains a place that has been built around the business of secrecy and national security, and which has lived through a series of transformations in those dynamics. While incomplete, it does offer a useful interpretive key, or point of entry, for studying US security and the reorganisation of domestic and foreign affairs in new forms of politics (biopolitics) emanating from nuclear technology and continuing through ongoing permutations of the US security state. The present paper represents the beginning of an attempt to consider some of these larger transformations side by side with specific ethnographic research about the reckonings of security in Los Alamos. While these issues are relevant for everyone and for the world at large, questions about them should be posed urgently in Los Alamos, whether or not they are on a routine basis.

NOTES

1. Of course, the encouragement and support of Islamic fundamentalism in strategic locations was itself a part of Cold War policy, and has been commented upon extensively.

2. Film *The Atomic City*, Paramount Pictures 1952. This film depicts one of the peculiarities of Los Alamos life: the way that residents take regular explosives tests in stride, hardly noticing them.

3. The present article draws on participant-observation, historical-archival, and comparative research carried out in and pertaining to Los Alamos since 1990. The participant-observation included working in four groups within the Laboratory, including Personnel Security, Classification, and Nuclear Technologies, as well as in sections of the Department of Energy and the Brookings Institution in Washington, DC, and holding a security clearance. Several hundred interviews have been conducted with scientists, engineers, technicians, support staff, and townspeople not directly employed by the Laboratory (teachers, shop owners, bus drivers, etc.). The interviews concern the time period of the Manhattan Project through the present. Unless otherwise noted all interviews were conducted in Los Alamos. Pseudonyms are used for all research informants as a consequence of anonymity agreements. While no informants discussed any classified material with me, it was often the case that they were more at ease talking anonymously, and felt they could be more candid about non-classified aspects of their work and experience. A further description of this longitudinal research project, the Center for the Ethnographic and Historical Research of Los Alamos and National Security, is available at <http://abmsc.org/LosAlamos.html>, accessed 20 July 2010.

4. Though Los Alamos is significant and of interest in and of itself given its historical association to national security, study of Los Alamos is further important for understanding how the business of security and military research permeates everyday social life – perhaps this is even more germane given the recent release of the *Washington Post* report on 'Top Secret America' which indicates that more than 850,000 Americans currently have Top Secret level clearances (presumably many more have Secret or other clearances, indicating that the security state permeates, to a significant degree, the fabric of American life) – <http://projects.washingtonpost.com/top-secret-america/articles/>, accessed 20 July 2010.

5. The strategy and tactics of the European theatre were discussed with me by Paul and by Warren, colonels in US Army Intelligence and Infantry, respectively, who had served extensively in West Germany. The same considerations about rendering the domestic space into a battlefield might well be applied to Korea as well, where a condition of war still technically obtains.

6. There is extensive literature on this. As important sources see Carole Gallagher's *American Ground Zero: The Secret Nuclear War* (Cambridge: MIT Press 1993), and Eileen Welsome's *The Plutonium Files: America's Secret Medical Experiments During the Cold War* (New York: Dial Press 1999).

7. Interview with Jamie, 16 March 2002.

8. In this respect the effects in Los Alamos could be useful in understanding current transformations in, for instance, San Antonio, Texas, which has been taking on a noted increase in intelligence work through the addition of new facilities.

9. This separation of clearances and domains within US security institutions is an artifact of the bureaucratisation of security functions within the government, and the attendant fact that different organisations have different (if sometimes overlapping) mandates, powers, and abilities. Each organisation is responsible for physical and personnel security on its sites, and each can hold its own classified information, which it 'owns' in the intelligence parlance, and to which it controls access. Thus the DOE 'Q Cleared' area is designed primarily around the access to DOE Restricted Data nuclear information for qualified persons (while it would be illegal to have this information outside the area), while the 'SCIF' is designed around access to CIA SCI information – although DOE and DOD SCI information is also dealt with in the SCIF. In practice information may cross the zones, but the general security architecture is designed around this principle of protected turf and access. The military services, the National Security Agency, and other government organisations also use the SCI Clearance, and 'own' their own intelligence. The SCI Clearance was formerly known as the SCSI (Sensitive Compartmented Secret Information) Clearance in some organisations.

Back of this system is the security tenet of compartmentalisation, in which the concentration of too much information and too many secrets in the hands of any one person is counter-acted by the breaking down of information, access, and projects to limit the comprehensive view of those within the security bureaucracy. Even within the same organisation among workers of the same clearance-type, information is typically broken down into discrete units requiring a 'Need to Know' for a specific work requirement to gain access. Clearly this can create its own problems, as some members within the bureaucracy need to be invested with wide-ranging access to information, making them possible liabilities. Also, too much compartmentalisation can impede scientific work and bring productivity to a standstill, especially in areas requiring exchange of ideas and information. Oppenheimer realised this and argued for a loosening of compartmentalisation among certain of the scientists of the Manhattan Project in order to promote exchange among them about pressing problems of the task, regardless of their speciality or group assignment. Without exception Manhattan Project scientists hold that certain issues could never have been surmounted without this cross-compartment exchange (and they say that it was common for 'non-experts' to contribute important solutions, i.e., a chemist would solve a problem that had vexed the physicists). Of course this may have left the Manhattan Project more vulnerable to espionage, as both Klaus Fuchs and George Koval took advantage of this de-compartmented access to compile varied and detailed information about several different aspects of the Manhattan Project, Fuchs of bomb design at Los Alamos and Koval about Uranium enrichment, processing, and metallurgy, as well as physics and design, from a number of sites of the Project.

10. A former Los Alamos employee who now works in the biotechnology industry explicitly compared the cultures of secrecy, while noting salient differences in terms of the national security versus profit motive logics underlying them, in a number of conversations with me from 6/2001 to 7/2010. Companies like Apple are notorious for their cultures of secrecy.

11. Interview with Nanette, 28 July 1994.

12. There is a sizable and growing literature on this important topic. Among some of the relevant works are V. Kuletz, *The Tainted Desert: Environmental and Social Ruin in the American West* (New York: Routledge 1997); W. Churchill, *Struggle for the Land: Native North American Resistance to Genocide, Ecocide, and Colonization* (San Francisco: City Lights 2002); D. Grinde and B. Johansson *Ecocide of Native America: Environmental Destruction of Indian Lands and Peoples* (Santa Fe: Clear Light Books 1998); N. Maclellan and J. Chesneaux, *After Moruroa: France in the South Pacific* (Melbourne: Ocean 1998); as well as B. and M.-T. Danielsson, *Moruroa, notre bombe coloniale: Histoire de la colonisation nucleaire de la Polynesie francaise* (Paris: Harmattan 1993).

13. See Grinde and Johansen, 'The High Cost of Uranium', in *Ecocide of Native America* (note 12) pp . 203–219.

14. Ibid., p. 218.

15. M. Foucault, *Sécurité, territoire, population: Cours au* Collège de France, 1977–1978 (Paris: Hautes Etudes 2004), pp. 7–10, 113.

16. Grinde and Johansen (note 12) p. 206. They note that including reservation and Indian treaty land the share of uranium on Indian lands within the US is 60 percent or more.

17. M. Foucault, *La volonté de savoir: Histoire de la sexualité I* (Paris: Gallimard 1976) p. 181.

18. G. Agamben, *Homo Sacer: Il potere sovrano e la nuda vita* (Torino: Einaudi 1995).

19. M. Foucault, *Il faut défendre la société* (Paris: Hautes Études 1997) p. 229. "Vous comprenez, par consequent, l'importance – j'allais dire l'importance vitale – du racisme dans l'exercice d'un tel pouvoir: c'est la condition sous laquelle on peut exercer le droit de tuer. Si le pouvoir de normalisation veut exercer le vieux droit souverain de tuer, il faut qu'il passe par le racisme. Et si, inversement, un pouvoir de souveraineté, c'est à dire un pouvoir qui a droit de vie et de mort, veut fonctionner avec les instruments, avec les mécanismes, avec la technologie de la normalisation, il faut qu'il passe lui aussi par le racisme. Bien entendu, par mise à mort je n'entends pas simplement le meurtre direct, mais aussi tout ce qui peut être meurtre indirect: le fait d'exposer à la mort, de multiplier pour certains le risque de mort ou, tout simplement, la mort politique, l'expulsion, le rejet, etc."

20. See D. Brugge, T. Benally, and E. Yazzie-Lewis (eds.), *The Navajo People and Uranium Mining* (Albuquerque: UNM Press 2007); and S. Udall, *The Myths of August: A Personal Exploration of our Tragic Cold War Affair with the Atom* (New York: Pantheon 1994). Udall represented Navajo workers and widows in a suit against the federal government over health effects from uranium mining.

21. Maclellan and Chesneaux (note 12) p. 94.

22. S. Graham, *Cities Under Siege: The New Military Urbanism* (London: Verso 2009).

23. L. Eden, *Whole World on Fire: Organizations, Knowledge, & Nuclear Weapons Devastation* (Ithaca: Cornell University Press 2004)

24. Foucault, *Il faut défendre la société* (note 19) p. 226.

25. J. Orr, *Panic Diaries: A Genealogy of Panic Disorder*, Durham, Duke University Press, 2006.

26. Foucault, *Il faut défendre la société* (note 19) p. 230. "Un racisme de la guerre, nouveau à la fin du XIXe siècle, et qui était, je crois, nécessité par le fait qu'un bio-pouvoir, quand il voulait faire la guerre, comment pouvait-il articuler et la volonté de détruire l'adversaire et le risque qu'il prenait de tuer ceux-la mêmes dont il devait, par définition, protéger, aménager, multiplier la vie?"

27. Ibid., pp. 232–233. "Ce n'est pas simplement la déstruction des autres races qui est l'objectif du régime nazi. La destruction des autres races est l'une des faces du projet, l'autre face étant d'exposer sa propre race au danger absolu et universel de la mort. Le risque de la mourir, l'exposition à la destruction totale, est un des principes inscrits parmi les devoirs fondamentaux de l'obéissance nazi, et parmi les

objectifs essentiels de la politique. Il faut qu'l'on arrive à un point tel que la population tout entière soit exposée à la mort. Seule cette exposition universelle de toute la population à la mort pourra effectivement la constituer comme race supérieure et la régénérer définitivement face aux races qui auront été totalement exterminées ou qui seront définitivement asservies."

28. M. Foucault, *Naissance de la biopolitique* (Paris: Hautes Études 2004).

29. For a brief geographic consideration of the role of the bomb in relation to population and territory, see 'Les territoires de nucléaires', available at <http://www.cafe-geo.net/article.php3?id_article=62>, accessed 20 July 2010.

30. Some sources which bear on this transformation are C. Rossiter, *Constitutional Dictatorship* (Princeton: Princeton University Press 1948); G. Agamben, *Stato di eccezione* (Torino: Bollati Boringhieri 2003); and J. Masco, 'The Nuclear State of Emergency', in his *The Nuclear Borderlands: the Manhattan Project in Post-Cold War New Mexico* (Princeton: Princeton University Press, 2006). G. Wills's recent *Bomb Power: The Modern Presidency and the National Security State* (New York: Penguin 2010) makes a similar argument. This transformation is particularly important to understand historically since the country seems to be again in such a period of the expansion and diffusion of the security apparatus after September 11. Again, the recent *Washington Post* report about this indicates the scale of the changes and the range of diffusion of security work throughout the nation. According to Foucault's concept of the *dispositif* as a tool for charting genealogical persistence and change, understanding this earlier, nuclear age, transformation of the US state is invaluable in coming to terms with the more recent, post–September 11th shift.

31. See K. Bird and M. Sherwin, *American Prometheus: The Triumph and Tragedy of J. Robert Oppenheimer* (New York: Random House 2006) pp. 210–212.

32. Of course this militarisation of the University of California has been the object of a good deal of contestation and activism within the system. Financially and structurally, though, the system has remained closely tied to defence science. Data about the collaboration of University of California scientists with Los Alamos, Livermore, and other National Laboratories is contained in many of the interviews conducted for this research.

33. For a powerful and sobering account of the reach of the security state in the nuclear age, see J. Masco's *The Nuclear Borderlands* (note 30).

34. As a description of the scale of this change, see D. J. Kevles's *The Physicists: The History of a Scientific Community in Modern America* (New York: Alfred A. Knopf 1995). He follows the sea-change in the position of physicists in American society around the event of World War II – from eccentric eggheads akin to philosophers earlier in the twentieth century to major agents of state power (and recipients of state funding) in and after World War II. In terms of general changes in US state structure see S. Lens, *Permanent War: The Militarization of America* (New York: Schocken 1987).

35. See for instance A. Kimball-Smith, *A Peril and a Hope: The Scientist's Movement in America, 1945–1947* (Cambridge: MIT Press 1970); S. Leslie, *The Cold War and American Science: The Military-Industrial-Academic Complex at MIT and Stanford* (New York: Columbia University Press 1993); J. Wang, *American Science in an Age of Anxiety: Scientists, Anticommunism, and the Cold War* (Chapel Hill, NC : University of North Carolina Press 1997)

36. See for instance L. Groves, *Now it Can Be Told: The Story of the Manhattan Project* (New York: Perseus, 1983).

37. Though there were of course some neighbours, and those evicted from the Pajarito Plateau to make room for the Army's Manhattan Engineering District Project including Spanish homesteader farmers (who self-identified as Spanish due to their lineage or connection to Spanish settlers in the area) and an elite boys' ranch school. Land was claimed from both the San Ildefonso and Santa Clara Pueblos, even though no physical removal of persons or habitations took place (although access to burial grounds and to ancestral ruins was disrupted). While each group had to engage in legal battles to gain compensation, the way was much easier for the rich and white agents of the Los Alamos Ranch School. The claims of the Spanish homesteaders and of the Pueblos were only settled after the turn of the millennium, with some still claiming the settlement inadequate (author interviews). See also J. D. Wirth and L. H. Aldrich, *Los Alamos: The Ranch School Years, 1917–1943* (Albuquerque: UNM Press 2003).

38. Investigation of the economy of power is the guiding question of Agamben's *Il Regno e la Gloria: Per una genealogia teologica dell' economia e del governo* (Milano: Neri Pozza 2007). It is also an abiding concern in Foucault's analysis in several sections of *Sécurité, territoire, population*, which are taken up by Agamben in his book.

39. Interview with Don, 28 June 1991.

40. R. Feynman, *Surely You're Joking Mr. Feynman* (New York: Norton 1985); and 'Los Alamos from Below', available at <http://calteches.library.caltech.edu/14/2/FeynmanLosAlamos.htm>, accessed 20 July 2010.

41. In interviews with those who had grown up in Los Alamos in the 12 years following World War II and before the Fence came down, flaunting or taking advantage of the Fence are frequent themes, especially among those who were teenage boys at the time.

42. This is a lesson which may have to be learned again, as recent reports indicate that some 900,000 Americans currently have Top Secret clearance.

43. Interview with Matt, 2 July 2006.

44. Interview with Roger, 16 Nov. 2005.

45. Interview with Sean, 15 March 2004.

46. One example of this is the SCIF discussed earlier in that paper.

47. "Pit" refers to the roughly spherical fission primary (usually made from plutonium) used to generate a larger thermonuclear reaction in hydrogen bombs. Interviews with TA-55 personnel Linda, James, Ruth and John, Nov. 2005 to Dec. 2009.

48. Interview with Juan, 16 June 1995.

49. Interviews with John, 9 Aug. 1996 and 10 July 2006.

50. Road diversions and the installation of physical barriers became a common security practice in the US after 9/11.

51. Interviews with Don, 8 July 1996 and 12 Oct. 1998.

52. In 2004 the National Nuclear Security Administration began to transition the operations of TA-18 to the Nevada Test Site and to other locations within Los Alamos National Laboratory.

53. Interview with Seth, 12 June 1994.

54. In a follow-up interview (2006) Seth commented how now people are well aware of this trajectory through organisations like Blackwater (now Xe), but that Los Alamos was the first private effort (a private security organisation under contract to the government) to take advantage of this flow of ex-special operations soldiers.

55. Interview with David, 14 June 1994.

56. Targeting is when population is "isolat[ed] [as] an objective," writes Derek Gregory. D. Gregory, "'In Another Time Zone, The Bombs Fall Unsafely'': Targets, Civilians and Late Modern War', *Arab World Geographer* 9/2 (2007) pp. 88–111.

57. See M. Bell Chambers, *The Battle for Civil Rights or How Los Alamos Became a County* (Los Alamos: Los Alamos Historical Society, Monograph # 3 in "The Story of Los Alamos" Series, 1999); and M. Bell Chambers and L. K. Aldrich, *Los Alamos, New Mexico, a Survey to 1949* (Los Alamos: Los Alamos Historical Society, Monograph # 1 in "The Story of Los Alamos" Series, 1999).

The Geographical Imaginations of Video Games: *Diplomacy, Civilization, America's Army* and *Grand Theft Auto IV*

MARK B. SALTER

School of Political Studies, University of Ottawa, Ontario, Canada

Video games are important sites for critical geopolitics, and this article engages in the analysis of Diplomacy, Civilization, America's Army, and Grand Theft Auto IV in order to understand how the geopolitical imaginary works in popular culture. It makes the argument that the claim to geopolitical and tactical verisimilitude is at odds with the representations of violence and the body in war. It concludes by mapping new directions for the study of video games.

Taking video games as a site of serious analysis for critical geopolitics allows us to reflect on social and cultural processes of militarisation and the construction and contestation of the popular international geographical imaginary.[1] In particular, we should apply the same kinds of analysis to video games that movies have enjoyed.[2] Power's brilliant recent article uses a socio-cultural method to study "the increasingly close relationship between the US military and the video-game industry, and the contribution that this has made to the militarization of US popular culture", because games are artifacts through which "geopolitical sensibilities emerge".[3] In this article I want to push forward this kind of analysis of gaming as a text of critical geopolitics and critical security studies.[4] I argue that in a wide variety of games, there is a coding of both territory and violence that reflects new dynamics in contemporary war fighting – that is the simultaneous disappearance and reappearance of bodies, and a similar disappearance and reappearance of "real" places of combat. *Diplomacy* and the *Civilization* series, both wildly popular, represent two dominant ways of representing space, territory and control from the Archimedean, top-down perspective. While *Diplomacy* is tied to a particular and familiar geopolitical context of nineteenth- century, pre–World War I Europe and *Civilization* allows for the deep customisation of its worlds, they both share some core geopolitical assumptions about territory, control, contiguity and conflict – in essence, they reproduce the territorial trap.[5] *America's Army* (*AA*) and *Grand Theft Auto IV: Liberty City* (*GTA IV*) best illustrate these corporeal dynamics as leading games in their genres. *AA* is the official game of the US Army, whereas *GTA IV* is a cultural phenomenon, rivalling summer blockbusters in sales and market-share. What differentiates the game dynamics of *Civilization*, *America's Army* and *Grand Theft Auto* from other possible games is the 'freedom' to win: both are classified as 'sandbox' games[6] because players can make what experience they want out of the material in the game. While the sandbox in *AA* is considerably smaller than that of *GTA:IV* and Liberty City smaller than the possible worlds of *Civilization*, there is the same conceit of freedom and authenticity. One blogger writes, "Sandbox games let me take control, find sense, or, if I must, create madness. These games offer choice . . .".[7] Freedom, in particular, the freedom to do morally, legally, or socially prohibited activities, is "the ultimate promise of so-called new media: virtual reality, the internet and videogames . . ."[8] In particular, I want to highlight some of the game dynamics that decorporealise the gamer-subject-citizen while rendering the target-object-victim into body parts, simultaneously invoking a verisimilitude of geopolitical conflicts while rendering the actual territories of conflict invisible. War games represent a militaristic, masculinist, Western geopolitical frame of violence. One of the ways that this is achieved is precisely by a differential corporealisation: Western bodies are abstracted and continually respawned – target bodies are disassembled and disappear. This has the effect of depoliticising the violence inherent in war and indicates a fundamental shift in the conception of the body and the (virtual) battlefield: simultaneously making verisimilitude a feature of the game and undermining those effects of violence in the name of verisimilitude. The spaces of both geopolitical and urban conflict are defined in terms of the "real" but in ways that undermine other representations of the real.

One of the tricks of modern war to make both bodies and space disappear — the body politic of the state is transformed, the bodies of soldiers are obscured, enemies rendered into body parts; space is elided into economic networks, cyberspace, speed, and in Augé's term "non-places."[9] This has deep roots in the Western tradition, as Coker argues: the focus of the warrior ethic is both a sacrifice of and a transcendence of the bodily.[10] Der Derian makes this argument strongly in terms of space throughout his early work: territory has been undone by speed.[11] Yet, as Scarry contends, it is precisely the body and the sacrament of that corporeal sacrifice that yields war as the ultimate method of dispute arbitration:

> there is no advantage to settling an international dispute by means of war rather than by a song contest or a chess game except that in the moment

when the contestants step out of the song contest, it is immediately apparent that the outcome was arrived at by a series of rules that were agreed to and that can now be disagreed to, a series of rules whose force of reality cannot survive the end of the contest because that reality was brought about by human acts of participation and is dispelled when the participation ceases. The rules of war are equally arbitrary and again depend on convention, agreement, and participation; but the legitimacy of the outcome outlives the end of the contest because so many of its participants are frozen in a permanent act of participation: that is, the winning issue or ideology achieves for a time the force and status of material 'fact' by the sheer material weight of the multitudes of damaged and opened human bodies.[12]

It is the double-bind of war on the body to rely upon the sacrificial body and deny the body has been sacrificed. Two dominant arguments in the field of critical security studies on the importance of popular culture are useful for critical geopolitics. Hansen (2006), Shapiro and Weber, for example, use documents of popular culture such as movies or cartoons, to understand the discourses and ideologies of masculinist, racist, nationalist, cartographical imaginaries of global politics. Shapiro prefers "an anthropological rather than a strategic approach to war . . . [that] is more concerned with the interpretive practices that sustain the antagonistic predicates of war."[13] Thus, this kind of analysis is most often focused on self/other dichotomies which are the engine of conflict. Der Derian and Coker, on the other hand, to make strange bedfellows, argue that the activity of war itself has become transformed through the social technologies of warfare and representation, to represent war as hyperreal or managerial it seems to me makes the same assumption about the character of war. Gieselmann thus argues that "by presenting the game war in the same way as the realm war on television and the real war on television like a war in a computer game, both worlds become aesthetically married to each other. The real war appears like a game – without any human victims. The game appears to be as thrilling as a real war."[14] Coker goes further to emphasise the public discontent with the production values of actual combat footage in comparison to their gamic representations: "The resemblance between the computerized war games played by the public as entertainment and the computerized film of the missile attacks prompted some viewers to complain of the grainy quality and poor graphics of many of the images that appeared on their screens."[15] Methodologically, in an effort to think through war beyond the battlefield and in other spaces of war, this article reads video games as texts of critical geopolitics, paying close attention to two ways that Frasca defines as conveying a particular worldview: "in their rules (what can and cannot be done and what actions lead to which consequences), in their goals (what is rewarded and punished)"[16] and argues that the virtualisation of war tightens this double-bind of the war-body-politic through a comparison of *Diplomacy, Civilization, America's Army*, its accompanying the "Virtual Army Experience" (VAE), and *Grand Theft Auto IV: Liberty City*.

WAR GAMES: DO YOU WANT TO PLAY A GAME?

Following the cultural turn in both IR and critical geopolitics, a number of scholars have analysed elements of popular culture for their representation of global politics and war.[17] Der Derian, in particular, has written extensively about the use of simulation and war-gaming in the contemporary understandings of war, peace, and politics: "When war games and language games become practically indistinguishable, when the imitative, repetitive, and regressive powers of simulation negate any sense of original meaning, more than just peace is at risk . . . with increasing orders of verisimilitude, the simulations displayed a capability to precede and replace reality itself."[18] Excepting Powers and Dittmer, video games are largely underrepresented in scholarly analysis but are a large industry with a large population of players, distributed globally though concentrated in infrastructure-rich countries; they are also a clear window into the geopolitical imagination. In the American context, to take an obvious example, "Sixty-eight percent of American households play computer or video games; the average game player is 35 years old and has been playing games for 12 years; the average age of the most frequent game purchaser is 39 years old; forty percent of all game players are women (In fact, women over the age of 18 represent a significantly greater portion of the game-playing population (34%) than boys age 17 or younger (18%)).[19] This is not a complete international imagination, but one understood as "strongly gender coded scenarios of war, conquest and combat . . . amplified by the industry's ongoing negotiations with a base of young male hardcore fans."[20] The creation and support of the Frag Dolls by games publisher Ubisoft represents an interesting case here. They are "are a team of professional female gamers recruited by Ubisoft to promote their video games and represent the presence of women in the game industry. These gamer girls play and promote games at industry and game community events, compete in tournaments, and participate daily in online gamer geek activities. Started in 2004 by an open call for gamer girls with competitive gaming skills, the Frag Dolls immediately rocketed to the spotlight after winning the Rainbow Six 3: Black Arrow tournament in a shut-out at their debut tournament appearance."[21] Games are just as important, widespread, consumed, and telling as speculative fiction, such as Harry Potter or Star Trek. Games, as artifacts of culture, are more than simply mirrors of dominant ideologies or sites of sly resistance; they are also technologies of the self – particularly indicators of the ideals and limits of socially acceptable geopolitical behaviour. As Leonard writes, games offer insight into dominant ideologies, as well as the deployment of race, gender, and nationalism. From the privacy of one's home, game players are able to transport themselves into foreign and dangerous environments, often gaining pleasure through domination and control of weaker characters of color. Video games thus operate as a sophisticated commodity that plays on the desire of individuals to experience the other, breaking down real boundaries between 'communities' through virtual play, while simultaneously "teaching" its players about stereotypes, United States foreign policy, and legitimization of the status quo, to name only a few.[22]

While games can be "realist or fantastical,"[23] according to Galloway, the importance of the game is not dependent on its genre. Galloway continues, "in their very core, videogames do nothing but present contemporary political realities in relatively unmediated form. They solve the problem of political control, not by sublimating it as does the cinema, but by *making it coterminous with the entire game*, and in this way video games achieve a unique type of political transparency."[24] FPS games can also be seen as a clear mirror to contemporary politics in the rise of the anti-terror genre, such as the *SWAT*, *Rainbow Six* (both of which predate 9/11), *Ghost Recon* or *SOCOM: US Navy Seals* series. These are tactical or squad shooters and so rely on coordination of multiple avatars in counter-terror stealth and other hostage-rescue scenarios. While some games allow players to choose the 'terrorist' side, such as *Counter-Strike* (which also predates 9/11), the majority align the player with an American or coalition counter-terrorism elite team. In addition to relying on the common trope of Islamist or drug/people smuggling terrorists, we see the construction of 'terrorist' as abstract, apolitical, simply enemies. The resurgence of World War II games in the past 10 years also enables the replaying of a war in which the moral categories are clear and made clearer by a selective retelling of the conflict.[25] Parallels from these well-known, simplified narratives, then are mapped onto contemporary conflicts such as the 9/11 attacks being represented as a new Pearl Harbor.[26] "The games recreate, or simulate, a world, mostly Europe, as it was more than 60 years ago – a time when, if we follow the line of Baudrillard's thoughts, life was more close to life, more real than today."[27] The potential tropes for permissible enemies are a gauge on contemporary identity-games – fighting against insurgents, terrorists, criminals, Nazi's, petty thugs, etc., is justifiable (and fun), whereas fighting Americans or Israelis is not.

In addition to reading video games as a 'straight' mirror of contemporary spatial strategies (of warfare, empire, or colonialism), we can also understand video games to demonstrate the political relation of bodies to spaces. The first way that games relate bodies to spaces is through the configuration of the screen. The in-game view of the first-person shooter (FPS) is characterised by the game world occupying the entire field of vision, with the avatar's outstretched, often weaponised, hands rising from the bottom of the screen, giving the first perspective, with various HUD (heads-up display) configurations. The frame of game player's perspective is a crucial part of the way that a game means [with apologies to Robert Frost, as opposed to *what* a game means].[28]

Following postmodern analysis of scopophilia, we should not be surprised that the viewpoint of the game player is often described as a camera. In third-person perspective games (such as tactical shooters in which the player sees the entire avatar) or God perspective games (such as RTS games where the player sees the entire game grid – barring 'fog of war' effects), the screen is separated from the corporeal. *America's Army* is an FPS (albeit one that requires tactical cooperation) that relies on a first-person perspective. While players may model their (male) avatars before the game starts, once within the game, the players see the avatar's wrists and weapon, and some specialised screens for weapons technologies such as sniper scopes and remote weapons. Critical information is displayed on the HUD (heads up display), such as current weapon, position, objectives, contextual orders, and health. In many games, such as *AA*, the map is integrated into the setting of the game (i.e., accessed through a GPS-like machine, rolled out as a map, or seen as part of the HUD).

Other games are played from the "god" perspective: RTS series such as *Command and Conquer*, *Total War*, *Supreme Commander*, *Age of Empires*, or the incredibly popular *Civilization* franchise, each have camera perspectives that zoom from the individual level through to the world or global level. The seemless tracking of the zoom camera not only situates bodies within a field, but also grants the player a position ex deus from which to be able to see and to control all (or know precisely the limits of the fog of war). The representation of this Archimedean perspective plays into familiar tropes on the ordering of space through visuality. I examine two popular 'top-down' games below.

THE TOP-DOWN WORLD: DIPLOMACY *AND* CIVILIZATION

The board game *Diplomacy*, and its online equivalents, portray Europe in 1900: the victory condition is a single country controlling half of the board.[29] While the majority of *Diplomacy* is played in person or by mail, and has recently faded somewhat in popularity, "commercial board wargames are the genotype of video wargames . . . in particular early video wargames inherited from board wargames their game mechanics, settings, . . . and professional relations with the military."[30] First marketed in 1958 as *Realpolitik*, this game for seven players, each representing a Great Power, is balanced in the distribution of resources and the cartography of the board in such a way that no individual may win the game and gain territory without the active cooperation of others; however, one cannot win cooperatively. The game's

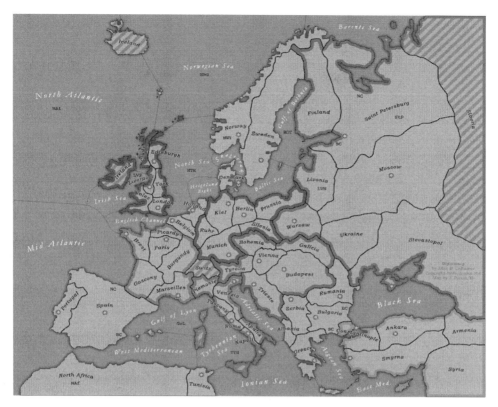

FIGURE 1 Diplomacy map (1999) (color figure available online).
Source: http://www.diplom.org/Online/maps.html

inventor, Allan Calhamer, specifically invoked MacKinder's conception of geopolitics in his description of the creation of the game.[31] The majority of gameplay, consequently, is devoted to making, breaking, and misrepresenting alliances in negotiation sessions: moves and support for conflict are revealed simultaneously. New pieces are awarded based on the capture of strategic spaces, which can be redistributed after which a new round begins. There are no rules for the negotiation phase, and so the only rules that govern the activity are social.[32] The geopolitical assumptions are relatively simple: armies advance and are supported – conflict is only possible at borders (see Figure 1).

Setting aside the nineteenth-century conception of war in which the victory in battles is determined by the number of units applied and supported (all units have the same weight, and only one unit may be housed in a province at a time), this is a Clausewitzian military-diplomatic system that particularly crafts the rules of military-diplomatic action to reinforce a zero-sum system, which reifies not only the dominance of sovereign states and their military capacity, but also, by continually referring to the absence of 'chance' in the rule book, affirms that foreign policy calculations are rational.[33] The anarchical, self-help, duplicitous nature of international relations as a distinct social realm of asociality is reinforced. As Dunningan states, there is a clear pedagogical function to this kind of wargame: it is "a historical account of an event in simulation form. The relentless organized approach of analytic history is unique. If you do it right, the [player] gets a large dollop of knowledge for a small investment of time."[34] There is a particular concretisation of the state (if not a direct praising of the balance of power system), because the state boundaries are immutable, the entire value of the country translated only into its military weight expressed in 'armies', and each move being one of contiguous, cross-boundary dispute or support. The world of *Diplomacy* is one entirely robbed of economy, culture, identity: geopolitics is understood exclusively in militaristic terms where states and state boundaries are fixed and conflict or cooperation arises only along borders.

Sid Meier's *Civilization*, originally marketed in 1991with sequels continuing to the present, takes a much larger scope: the entire globe (and indeed newly created worlds to escape the geopolitical constraints of Earth).[35] Like *Diplomacy*, players are matched against seven other civilisations – either actual players online or computer simulations. It allows for multiple victory conditions: a player starts with a single settlement and may win through military, economic, cultural, or technological domination (space race, in which one essentially exits the playing grid). As a 'civilization' (which have culturally specific advantages) along with a 'great leader' – such as Cleopatra of the Egyptians, Caesar of the Romans, Churchill of the British, players can make alliances and treaties in all military and economic domains, while cultural influence (such as the discovery or construction of a Wonder) is a field effect that can even induce conversion and political control. These categories "transpose the many layered quality of social life to an inflexible, reductive algorithm for 'civilization'."[36] Players start as despots, but later research and choose their styles of government, which confers advantages and costs, managing their citizens' happiness through the production of culture, food, and luxuries. In *Civilization*, we might argue that an international society exists in the English School sense, although, similar to *Diplomacy*, there are no inherent moral constraints on behaviour within the game, simply rational calculations to make. The teleological evolution of history brings different technological advances and ecological costs, causing different kinds of play in barbaric, classical, industrial, and modern ages.[37] Allowing for greater flexibility of limits, players may set the rules for their game, including the number and bellicosity of barbarians, the fixity or fluidity of alliances, and the possible domains of victory: while boundaries play an important role in conflict, trade, and cultural influence, there is more of a city-focus than a boundary-focus. Players manage cities, technology trees, diplomacy, infrastructure construction, and war in order to approach their world domination. Because resources are scarce and the technology tree the same for all players, "rush" strategies (also called "Mongol" strategy), are occasionally advocated in the game literature where a high number of low-tech fighting units are produced, eschewing invention, and sweep across the map to dominate space and therefore the game.[38] The focus on cultural Wonders and their disproportionate effects in later iterations (*Civ II-IV*) made a winning strategy harder to fathom. The geography of the game is chosen by the player: climate, amount of land, archipelago vs. continent, etc. But the fundamental structure of the control of space remains constant: cities produce a field effect "civilization" that is directly related to production, culture, and infrastructure. Thus, boundaries, borders, and adjacency no longer play the same role in *Civilization* as they do in *Diplomacy* – but the idea of single, unitary fields of spatial control dominated by urban centres still dominates the game-space.

One of the core concerns of IR theory and critical geopolitics is the justification and understanding of violence within the (international) realm. As Walker argues, it is one of the fundamental tricks of the discourse of the sovereign state to make the inside/outside boundary plot onto the acceptability of violence: safety within, violence without.[39] As a popular representation of the limits of violence, video games illustrate the complex politics between the violence and the bodily. I do not want to engage in the argument that video game violence promotes or fosters or legitimises *real* violence, precisely because represented violence is real and we must analyse it as a thing itself worthy of serious study.[40] Some games in this genre, such as *Soldier of Fortune*, make a selling point of viscera and the specificity of violence – although the player's avatar is most frequently abstract.[41] As both the *Civilization* and *Diplomacy* games claim, in their own way, to be realistic, other games can also re-inscribe of geopolitical imaginaries and narratives of conflict through their claims to authenticity by using real-world footage, surveillance photos and battlefield intelligence, actual conflicts, and technical data (particularly about weaponry). Setting aside for the moment *America's Army*, *Special Force*, or other games that attempt to intervene directly in the perception management of 'soldiering' or 'resistance,' a significant number of games stake their claim on the grounds of authenticity and historical accuracy. FPS games are characterised by "the subjective camera perspective, coupled with a weapon in the foreground."[42] The in-game narrative, in particular the back-story to the player's character, often gives us a sense of the possible – what international conflicts or storylines 'make sense.' *First to Fight* (2005), for example, a game 'based on the training tool used by the US Marines' re-places the player in modern day Beirut, where the Marines fight urban operations – which is both dependent upon and plays with the 1982 attack on the Marine barracks in Beirut by Islamic terrorists. *Frontlines: Fuel of War* (2008) is based on a dystopian 2024 in which the world has bifurcated into a new bipolarity (US + EU vs. Russia + China) over disappearing oil and the resultant energy crisis. *Special Force*, on the other hand, is a FPS shooter by Hizbullah's Central Internet Bureau that "allows the gamer to virtually participate in operations of the Islamic Resistance in South Lebanon."[43] As with movies, cartoons, novels, or other cultural artifacts, we can gain critical insight as to how compelling identities, scripts, and narratives are constructed through the genre of war games. In the initial sense, then, games are excellent source material for understanding the valourisation of militaristic interventions that elide historical contexts and ethical dilemmas, and also for demonstrating the limits of possible speech.[44] Games then can provide primary texts for analysing the *frame* for understanding global politics.

AMERICA'S ARMY: EMPIRE IN GOD MODE

America's Army is a free online, first-player tactical shooter. First launched in 2002, and now offering its third version, *AA* is an online recruitment tool and so can be understood as a technology of the soldier-citizen – a way for the US government to construct a particular virtual American subjectivity that recognises itself as part of a general war machine even as it abstracts from the corporeal reality of violence. *AA* bills itself as having "penetrated contemporary culture . . ." and as having "one of the most recognizable game brands"[45] in large part because it offers the "most authentic military experience available."[46] Nichols reports that "by 2005, the game had tallied more than 4.6 million registered users and was attracting roughly one hundred thousand players each month. Perhaps more significantly, more than 30 percent of Americans between the ages of 16 and 34 said that what they knew about the Army came from this video game."[47] Real soldiers and players discuss tactical and technical issues in the large online community in a way that further reinforces the claim of the game to be real. This public relations exercise is supplemented by "real heroes" – decorated veterans who now contribute to the *AA* experience, both the VAE and the *AA* online community – who are all men whose narratives of heroism, for which they have been awarded medals, reinforce a disregard of bodily injury.[48] The bodies of war in this game are entirely masculine: avatars may only be male (although skin colour may be adapted by the player). The female body is remarkably absent in the game[49], as is the misogynist discourse of military masculinity that is such a crucial part of representations of Army training.[50] In this section, I want to highlight two important aspects to the coding of the *AA* experience: the representation of self/others and its travelling show, the Virtual Army Experience.

AA's missions are played as a single player training scenarios or cooperative online battles, and involve the accomplishment of specific military goals (the capture or defence of intelligence, VIPs, or the conquest/defence of a strategic space or asset) within a pre-described set of rules of engagement. In this way, *AA* is a "top-down" shooter, far more over-coded with rules than most.[51] It offers both "honor service," which requires completion of training and rewards additional qualification with new missions with the ultimate goal of becoming a Green Beret, and "free play" or "instant action," which limits the amount of honour (traditionally called 'experience' in role playing games) that one can accrue to the character but allows access to missions throughout the programme. As the *AA* development team explains:

> In *AA* you earn access to online play by paying your dues in basic training (thus experiencing the Army's merit-based promotion) and qualify for good stuff like marksman, airborne, and medic through advanced classes. Basic teaches you to think Army-style and provides a handy space for learning how to maneuver before joining online play. **The very pace of play, which is deliberate compared with other action games, reminds the player that the Army proper is not a game**.[52]

Honour accrues only with the success of missions played by the ROEs (rules of engagement). As Bogost suggests, "The correlation of honor with the performance of arbitrary and decontextualized missions offers particular insight into the social reality of the U.S. Army".[53] The programmers are particularly engaging verisimilitude: the play reinforces that the Army is not game – thus *America's Army* undermines the claim to authenticity of other games, and I would argue the claim to authenticity of other representations of real combat.

AA matches are played online, either as a single player or in groups. All "others" are seen as enemies. All players engage the scenario as Americans: whether assault or defence the self is always American, the other is always terrorist.[54] This is a particularly neat political under-coding of war. "In *America's Army* the gamer always plays as a member of the U.S. Army, though they appear to the opponent as the enemy. . . . mirroring the rhetorical ways in which news reports [and others] wield 'we' and 'us' to conflate the complex logic of war into the more streamlined ideology of good versus evil and us versus them, the "we" and "us" in *America's Army* is always-already the U.S. Army."[55] One can never be a 'bad guy' in America's army, or rather, one is always already good to one's self and bad to the other: the moral coding is always complete. Colonel Casey Wardynski, the creator and project director for *America's Army* describes it this way: "Essentially both teams are Americas; so we can bind both teams to the value structures and rules of war and engagement the Army operates under. If we let kids elect to play 'bad guys' we couldn't bind them to the Army's rules and values, we could have mayhem like in an entertainment game, and the point would be lost."[56] For example, the briefing for the infantry mission presents both

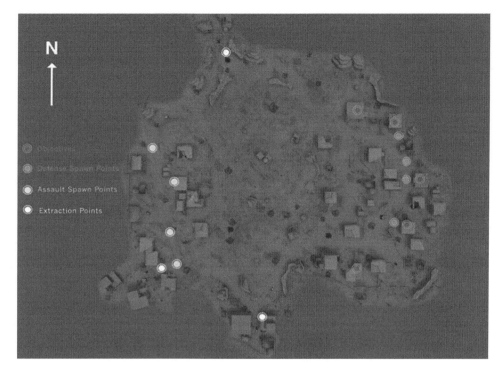

FIGURE 2 Border mission map (color figure available online).
Source: http://manual.americasarmy.com/index.php/Border_Tutorial, accessed 15 June 2010.

assault and defence missions as morally unambiguous, and as taking place in the same space (see Figure 2). "Border":

> *Assault*: A known terrorist cell has set up a base of operations in a small border village. Intelligence reports confirm they are planning a synchronized bombing strike at several different civilian landmarks. Intelligence also reports that high level sensitive information pertaining to this bombing strike is being kept at this location. Rules of Engagement: Use of force authorized against all positively identified enemy combatants. Be aware that there are other non-combatant personnel in the area. Minimize harm to local infrastructure.

> *Defense*: Having just secured important high level information within a remote border village, a defense must be set up to secure the village and the sensitive information contained within it. Intelligence reports that a terrorist counter-attack is imminent. Rules of Engagement: Use of force authorized against all positively identified enemy combatants. Be aware that there are other non-combatant personnel in the area. Minimize harm to local infrastructure.[57]

Throughout the missions, players are alerted to the presence 'non-combatants,' and the adherence to the laws of war and rules of engagement are described as minimising 'harm to local infrastructure.' This is a similar structure for the majority of missions: a combat scenario is provided in which players are always identified as Americans (sometimes protecting coalition forces, refugees, civilians, or humanitarian workers) and the enemies always described as terrorist, insurgent, or reactionaries. Nieborg explains the "software trick": "It is impossible to play as a terrorist and gamers never have to intentionally kill a US solider."[58] The players are thus always self-identified as being 'good guys' – and never shoot at other Americans. There are penalties for 'fragging' – or killing a fellow American soldier on the player's own team, which include a loss of honour and eventually incarceration in a virtual Leavenworth military prison. There are no penalties for tea-bagging, however, and other informal forms of hazing or humiliation. A common practice in *AA*, and other online multiplayer games, is the "teabagging" of fallen enemies – i.e., standing over a dead avatar and crouching repeatedly over the face. Once killed in multiplayer games, there is often a period of time before players "respawn" when they can see but cannot move, and so the fallen player sees scrotum of the killer on his screen repeatedly. Thus, teabagging is understood a form of heteronormative humiliation of the fallen enemy.[59]

The moral cartography of these games is entirely ego-centric: the self is always in the right, the positively identified enemy is always worthy of extermination. Two points on the actual ROEs within the missions. First, unlike many FPS, in which one is encouraged to shoot anything that moves (and often things that do not since the destructibility of the environment is a relatively new innovation), *AA* has a much tighter band of acceptable behaviour, or at least behaviour that will lead to the success of a mission. If a player does not play within the ROEs (including minimalising harm to infrastructure), the player does not gain honor (experience points) and cannot advance. Although, even with this overlay of military codes of conduct, given that the target audience for *AA* is the same pool of recruits it wishes to target for military service "it is no surprise then that games like *America's Army* aim to fulfill expectations common to a first-person shooter genre, even if it deviates from standard shooters in important ways by injecting doctrinal 'procedural rhetoric' into the gaming experience."[60] Second, as noted by a number of critics, the violence in *AA* is completely abstracted.[61] The lead designer is particularly coy about this fact: he says, "We don't downplay the fact that the Army *manages* violence."[62] Wounding the self is entirely iconic, represented by a drip of blood on the HUD (which can be healed by an in-game, specially trained teammate: CLS – combat life saver): "It is . . . impossible to bleed to death and the bleeding of an avatar is not graphically simulated. When a player looks at his hands, there is never blood on them nor do friendly or enemy casualties show any sign of injury."[63] This cleansing of the (virtual) battlefield extends to the enemy. "If you shoot an enemy, he merely sits down as if resting and after a while vanishes. You hear no death cries or see any dramatic animations – war is presented in a clean, almost sterile way."[64] Der Derian points to the absent casualties, both for enemy combatants and civilians as a crucial part of the *virtuous* war, by which he means a war that is both virtuous and virtual.[65] The claim to authenticity is based on "extremely detailed and accurate weapons and combat scenarios," however, the inevitable injury and death that results from these technologies and battlefields is markedly abstract.[66] Nichols again: "While the game has realistic sounds and sights, players who are killed hear no noise and are shown only a small red circle at the time of their virtual death. This design was done, in part, to ensure the game a Teen (T) rating."[67] While the training modules in *AA* attempt to translate the intense physical and psychological conditioning of recruits, the actual physical experience of players is simply the manipulation of mouse/keyboard or game controller. Fighting skills, fitness, and bodily experience are thus already abstracted and mediated by the game interface. But, the FPS frame and the structure of the health system in game assists this alienation: "The realities of carrying eighty-pound knapsacks in one hundred and twenty degree heat, the panic-inducing anxiety and fear of real people shooting real bullets or planting real bombs to kill or maim you and your fellow soldiers, and the months, if not years away from family are not among those experiences reproduced for instruction or entertainment."[68]

To be mindful of Lisle's claim to situate the actual production of cultural artifacts,[69] we must be mindful that *America's Army* was designed as a "strategic communication tool."[70] The online game is supplemented by the spectacle of the Virtual Army Experience, a travelling show that the US Army hosts at state fairs, air shows, and other similar public gatherings, launched in 2007 and continuing through 2010.[71] In the display, players sit in a military HMMWV and play the game on very large screens that display an urban rescue operation. According to its brand manager, "The VAE is the most effective recruitment tool the Army has today. Throughput averages 200/hr . . ."[72] Here again, we see the same differential corporealisation of self-other. As one blogger reports:

> I had fun during the intense but short experience. It felt surprisingly real, with the gun and Humvee shaking and rocking wildly as I shot at terrorists on a huge screen. Unfortunately, it didn't really present the same level of risk most video games offer. As far as I could tell, nobody in the simulation died or got hurt. Sure, bullets flew and bombs exploded, but nobody lost a life and had to respawn, or any other of the typical game conventions you'd expect from an FPS or a light-gun game. It was like I was playing through an Army mission in *god mode*. (emphasis added)[73]

God mode is a hack or a cheat in games which allows players to enjoy unlimited health, ammunition, and be invulnerable to attack: it is precisely the corporeal politics of *America's Army*. For this gamer, the *VAE* is distinguished from the *AA* mission precisely because of the lack of risk to the player's avatar – the lack of virtual damage to the abstract icon broke the veil of authenticity, and signalled the degree to which the actual physical experience of 'shaking and rocking wildly' that 'felt surprisingly real' was not real enough. The death of the terrorists goes completely unnoticed.

Just as the tactics of the games combat scenarios are abstract – enough to allow US soldiers to play either assault or defence roles – the geopolitics of *America's Army* is entirely abstract: the missions are not placed in any particular geographical space. This imaginative geography is similar to Der Derian's description of the Army National Training Center in the East Mojave Dessert.[74] *America's Army 3* takes place in Czervenia and the Republik Demokratzny za ta Ostregals. Google-like maps are offered to provide political, satellite and climate information (see Figure 3). The situation again provides a very clear narrative of moral superiority and reluctant military action: "U.N. Security Council resolutions failed to resolve the conflict and U.N. aid workers are overwhelmed. A humanitarian crisis of epic proportions is imminent if decisive action is not taken. The RDO government and the U.N. have requested the help of the United States. The President has sent the U.S. Army to resolve the situation."[75]

The wider regional scale is then eclipsed by particular mission environments, which each play host to a number of scenarios (escort, extraction, take and hold, etc): bridge; alley; impact; pipeline; ranch. Thus, the imaginative geography provides a tight moral justification for military action. Unlike *Diplomacy* or *Civilization* (or indeed other FPS shooters), the scale of mission is not broad, but rather very specific and precise. Unlike the *Call of Duty* or *Battlefield* series, for example, the player is not the linchpin of history whose actions determine the course of world history (and indeed, since mission success is linear, the player can only save the world at the end of the mission/game). Rather, missions and scenarios in *America's Army* are displayed as tactical, small-scale, local scene. While FPS games tied to a particular historical context must specify the enemy (Nazis, Japanese, Islamist), the 'other' in *AA* is always an enemy, terrorist, insurgent with no particular politics, cause, or grievance. One might read contemporary operations into the presence of 'noncombatant,' 'coalition,' and 'UN' forces – but arctic, jungle, and swamp missions are entirely absent of identifying data. Thus, the geopolitical imagination of the game enables the morality of Americans, supported by the unequivocal moral case for each particular mission or scenario. This is bolstered by the complete absence of the dead 'others': after a time of 'playing dead,' characters respawn.

In sum, the game is a "bold and reinforcement of American society and its positive moral perspective on military intervention, be it the war on terrorism or "shock and awe" in Iraq," and it reinforces a number of dominant geopolitical tropes about the American military role in empire.[76] The

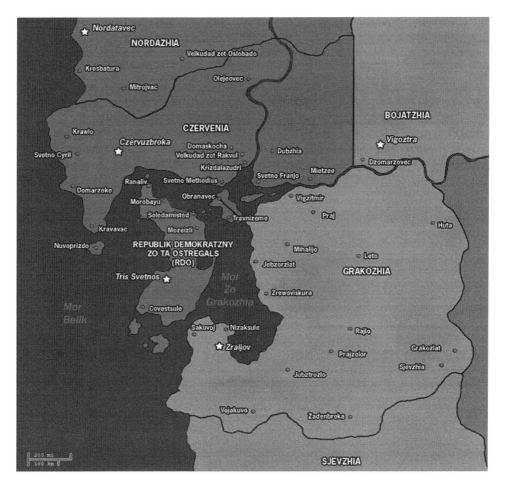

FIGURE 3 *America's Army 3* map (color figure available online).
Source: http://www.americasarmy.com/intel/recon.php

geopolitics of *AA* are entirely imaginary, and so each player is prepared for missions throughout the globe (except the United States). The abstract nature of the global war on terror and the geopolitics of the American empire both make the imperial centre a place of radical insecurity, as does the blurring of crime and terror, which is exemplified in the *Grand Theft Auto* series.

GTA IV: LIBERTY CITY

The launch of *Grand Theft Auto IV* (*GTA IV*) was a cultural event, and mirrored the release of a blockbuster movie: it grossed over $500M in the first week and sold over 3.6 million games in the first day.[77] In addition to being wildly popular, *GTA III* was the best selling video game of 2001, *Vice City* of 2002, *IV* of 2008, the series "is one of the most dominant media franchises of the new millennium and a cornerstone media point for millions of today's youth."[78] The GTA franchise (starting with *GTA III*), published by Rockstar Games, has essentially created and popularised the first-person "sandbox" genre of game that involves a large-scale world, living and breathing independently of the player character, in which a certain linear mission tree is possible – but players are "free" to interact with the game world entirely how they please, described as non-linear play. The only playable character in *GTA IV* is Nico Bellic (an immigrant), and while there are female non-playable characters (NPCs), there is no customisation available on Nico. Within missions, clear objectives and goals circumscribe the action, but outside of missions, characters may interact with the virtual world however they please with much less constraining rules: drive a taxi for fares, pick up the wounded in an ambulance and deliver them to hospital, or access a police computer and conduct vigilante attacks. One of the stand-outs for *GTA:IV* is the dialogue, sounds, and sheer interactivity of the world: if a player nudges a pedestrian with a car, that pedestrian will respond (hey, I'm walkin' here); or phone the police on their cell if they see a crime; or run away if a weapon is pointed at them; police will call for back-up; taxi-drivers will make chit-chat; the radio news will includes stories of the players recent crimes. These minutiae of quotidian life all enhance the game's claim to authenticity, even as the very premise (and the behaviour it promotes) far exceeds the normal bounds of society.

As the title would imply, stealing cars (boats, helicopters, motorcycles, fire trucks, ambulances, etc.) is encouraged – and the (vehicular) killing of the many pedestrians is only slightly discouraged (in earlier versions, these acts were encouraged and rewarded).[79] This is because the outer-limits of the morality in *GTA IV* are described by a police algorithm: the character accrues stars that correlate to his "wanted" status – stealing a car in view of a police officer brings one star, while destroying a police car (with a grenade or Molotov cocktail, for example) brings three stars. All crime incurs the interest of the police, but these stars correlate with the degree and gravity of police response, and relate to a particular search radius around the character, within which police will take on pursuit and/or engagement. The higher the star rating, the wider the diameter of the area, which flashes blue and red on the mini-map visible on the HUD. Thus, geographically, the space of security is always centred around the player character, independent of the area or neighbourhood. And, because the limits of police interest/capacity are clearly visible (promoting perhaps a managerial view of the limits of modern policing?) the player is always encouraged to (try to) escape.[80]

As with all of the *GTA* games, relationships (be it criminal, familial, or sexual) play a key role. However, the core relationships have been widely and roundly criticised for their repetition (if not direct reinforcement) of racial and sexual stereotypes. Amongst the most grievous, visits to sex workers replenish the character's health and in *GTA: IV* strip clubs play a crucial role in the narrative. Clearly, some degree of notoriety is useful from a marketing point of view. *Vice City* included in its first iteration a mission titled "Kill all the Haitians," which was amended in later releases after a public outcry. *San Andreas* included programming that could be unlocked with a hack titled "Hot Coffee" that allowed the player to engage in (clothed) avatar sex – which was subsequently patched by Rockstar. The racial dynamics of the American imaginary are in full play – mediated through popular culture (both in their production by the Scottish developers and in their repetition in the game narrative and references). The playable character in *GTA III* is a white male; *Vice City* an Italian-American; *San Andreas* a young African American; *GTA: IV* a middle-aged Eastern European. It can be argued that "by portraying white characters as being smart and powerful . . . authoritative (the police) or as the main character with whom players must identify – the discourse in GTA3 implies typical notions of white supremacy."[81] Similarly *San Andreas* focuses on an African American player character and a 1990s rap-inspired plot line: "There is a glamorizing, and even spectacularization of violence, a marking of young black bodies as disposable, and insistence on a culture of cynicism as well as a particular formation of African-American experience that is extremely problematic."[82] However, DeVane and Squire demonstrate through their field research that these racial stereotypes are interpreted differently by diverse groups of gamers: "Peripheral social groups within the dominant class – white working- or middle-class – enjoyed the satire of GTA: San Andreas but displayed concern about stereotypical representations of race. Conversely, participants from socially and economically marginalized groups – African American, working-class, or working poor – used the game as a framework to discuss institutional racism in society."[83] Whether the satire reinforces contemporary stereotypes and dynamics of socio-economic and political power, or through humour "the player is provided with cues for reflecting and evaluating his/her own perspectives on issues of race," these racial and gendered dynamics circulate prodigiously within the game space.[84] The interpretation of these signs is crucial: "There is a fine line to (t)read between parodic critique and discursive reinscription, especially in relation to the deployment of racialized archetypes and the persistent linkage of these archetypes with criminal elements."[85]

The *GTA* games have become a clear mirror of American geopolitical and particularly urban imaginaries.[86] GTA *III*, released in October 2001 – and thus in production long before 9/11 – was set in the fictional Liberty City, based on New York City. Its plot involves a would-be bank robber who is shot by his girlfriend and accomplice, left for dead, arrested and tried for the crime, but escapes and rebuilds a life in the criminal underworld. With some loose parallels to the *Godfather*, this game plays with familiar conceptions of organised crime. This is a pre-9/11 game with a distinctly local world. *GTA: Vice City* was released in 2002 and set in Miami – it used the war on drugs in the 1980s as its context indicated by in-game fashion, music, and media. The player character is an ex-con Mafioso who is sent to Vice City from Liberty City to run their cocaine operation; he slowly climbs the ladder of the drug underworld and, after a *Scarface*-inspired showdown, defeats his former patrons to become the head of his own syndicate. The world of *Vice City* is more international, because it is implicated and involved in the circuit of illegal drugs and US government's war on drugs. *GTA: San Andreas* is set in 1990s California, released in 2004, offers correlates of Los Angeles, San Francisco, and Las Vegas with a much narrower frame of a stereotypical African American urban life: police corruption, gang warfare, drugs, and rap music. The Rockstar Games programmers were not familiar with the 'real' California/Nevada. Rather, the game space "is an oddly global artifact, the result of a team of Scottish developers raised with the Los Angeles depicted in N.W.A. music and Spike Lee films exporting that culture back to Americans."[87] The plot hinges on police and judicial corruption, culminating in a city-wide riot (similar to the 1992 Rodney King riots, fitting with the historical frame of the game). In this case, the NPC population is

> represented in *San Andreas* . . . as a site of terror, insecurity, and uncertainty. The public arena is marked as a site where violence is not only probable, but imminent. Death occurs absolutely meaninglessly and indiscriminately. If one could shift the narrative focus of the game from [the player character] to one of the civilians . . . the game would be nothing more than a countdown to a random, violent death.[88]

Notably, *San Andreas* included a much wider game area (thirty-six square kilometers by the scale of the game), an interactive non-playable character (NPC) population (that obey traffic signs, etc.), and a different conception of American space than the previous *GTA* series: major cities were separated by vast areas of desert. Here is an urban America that is obsessed with the repossession of what it has lost, the incompetence or corruption of its police, and an environment that is focused on cities situated in a hostile, empty desert.

GTA:IV returns to a Liberty City that is much more faithful to New York City, including the boroughs of Queens, Brooklyn, Manhattan, the Bronx, and New Jersey (see Figure 4).

The main character, Nico Bellic (a clear reference to Niccolò Machiavelli and *belli*, the Latin root for war), does not originate from Liberty City, but rather comes to Liberty City in search of the American dream – much hyped by his cousin Roman (although he is also fleeing his former criminal connections in the Adriatic). Nico has been people-smuggling, but is chiefly portrayed as a veteran of the Bosnian wars, searching for a former army

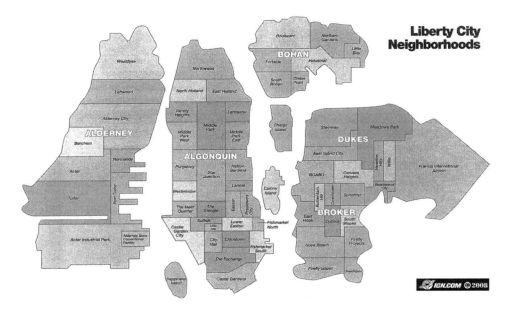

FIGURE 4 Liberty City neighbourhoods (color figure available online).
Source: http://grandtheftauto.ign.com/wiki/Image: GTA4–Liberty-City-Neighborhoods3.jpg, accessed 16 June 2010.

colleague who 'betrayed' his unit: there are intimations that Nico "did bad things" – i.e., committed atrocities – as a soldier, which is why the moral bounds of his behaviour have been loosened, a fact lamented by his mother in e-mail exchanges. In previous *GTA* games, criminality and brutality have been justified because of the internal socio-economic logics of the criminal underworld (and largely outside of the protagonist's control). In *GTA: IV*, Nico's violent and anti-social actions are simply the result of an external condition of 'war' and his disillusionment with the American dream. The *GTA* franchise is well-known for its satire of American popular culture; the inscription on the Statue of Happiness seems particularly stinging: "*Send us your brightest, your smartest, your most intelligent, Yearning to breathe free and submit to our authority, Watch us trick them into wiping rich people's asses, While we convince them it's a land of opportunity.*"

GTA: IV also reflects a new spatial imaginary. Liberty City, whose motto is "the Worst Place in America," reflects much more closely the actual New York City, complete with a Statue of Happiness, Middle Park, and Grand Easton Terminal, which is in part a result of improved geomatic and graphics technology. The development team at Microsoft's Xbox headquarters were playing online — and one native New Yorker stole a bus, invited all avatars aboard, and gave his fellow gamers a virtual tour of the 'real' city. The developers used "time-lapse video recordings to monitor traffic patterns and rainfall, photographs of more than 100,000 locations, and regular site visits to investigate the ethnic 'character' of different corners of the city."[89] Again, the recurring theme of verisimilitude plays a key role in the gaming experience. The empty spaces of *GTA: San Andreas* have disappeared; however, the war on terror appears as absolute spatial limit on the game space. In the *GTA* franchise, players are initially restricted to one small area, gaining access to the entire game space as they complete the linear missions. In previous editions, the bridges, highways, and tunnels that separate the areas were closed due to maintenance of America's crumbling infrastructure. However, in *GTA: IV*, those areas are closed off by police because of "terrorism." Thus, the limit of the playable area of the sandbox game is literally described through the war on terror. It is possible in the game to get around the barricade by stealing a boat or helicopter or to simply shoot one's way through, but this causes a six star wanted level and an immediate and massive military response led by the NOOSE (National Office Of Security Enforcement) with Apache-style military helicopters. When in these terrorised zones, there are no NPCs (a hallmark of the franchise). The barricades are removed once the linear missions are completed. Following work by Graham,[90] I want to argue that the focus on New York City as a localised site of the global war on terror demonstrates the reorientation of the American military imaginary towards the domination of urban space and surveillance.[91]

The endgame is also revealing of new realities (spoiler alert). Nico completes missions including assassinations for a shady government department (United Liberty Paper), which, we find out, has been stalking him since the beginning of the game: his first 'girl friend' Michelle was a 'honey trap,' and gains knowledge enough of his crimes on American soil to blackmail him into working for this agency. As a reward, the UL Paper arranges to have Darko Brevic, the target of Nico's revenge fantasy, brought to Liberty City from Bucharest. *GTA: IV* offers several 'choices' – where the player may either execute or spare an important NPC (with different consequences for each decision). Two finales are possible: both of which lead to the death of an NPC with whom Nico has a long-term and serious relationship (cousin or girlfriend). The incorporation of substantial elements of the global war on terror reproduces a new American geopolitical imaginary: the acceptance of the division of urban space into terrorised and policed, extraordinary rendition, government agencies which operate outside the law or judicial oversight.

The visual frame of *GTA IV* is a third-person action/driving game (not an FPS). The game's camera hangs above and behind the main character, which can be rotated along certain axes. There are three different possible perspectives, including a 'cinematic' view in which the central character is the object of distant panning shots. Consequently, *GTA IV* does not make a visual equation between the viewpoint of the player and the viewpoint of the character.[92] *GTA: IV* steps back from *GTA III* in terms of the physical malleability of the character. *GTA III* included a sub-system by which the character's body could be changed to be more muscular, fit, or overweight through exercise and diet.[93] Nico in *GTA: IV* is a fixed, physical character. As in the *America's Army* game, health is represented by a red circle on the HUD – and health is completely abstract. Unlike *AA*, where health cannot be replenished within a mission (except by a CLS), *GTA: IV* offers numerous ways for Nico to replenish his health: eating a burger, drinking a soda, calling the paramedics, visiting a prostitute, or going to sleep. When shot, blood spatters briefly on the camera lens, but no damage is seen to Nico's body. This may indicate a distancing between the violence experienced by the character and the violence processed by the character. DeVane and Squire report that a gamer said that the violence was "less influential because it's a third-person game and not a first-person shooter. Because of the angle . . . it's like the angle . . . it's different."[94] When in a physical accident (traffic, base jumping, falling from a helicopter, etc.), there is a small blood spatter on the point of impact or pavement which quickly fades. If Nico's health is completely depleted, he is taken to one of several hospitals, and is charged 10% of his total wealth (up to a total of $10,000): none of his weapons are confiscated. This is a fascinating elision of the politics of healthcare in the United States and presents the image of a system built on a flat-tax basis. The (temporary) death of Nico in *GTA: IV* is always replayed cinematically in slow-motion with the use of 'ragdoll' physics (although mission-critical NPCs cannot be killed before fulfilling their narrative function). Following the common convention, the dead NPCs disappear.

The *Grand Theft Auto* franchise is not a cognate of *America's Army* as a direct propaganda tool, but does contribute to our critical understanding of the new militarism within popular culture. Nico Bellic's narrative is much more implicated than previous GTA characters in world politics, which points to a new geopolitical imaginary that is not only local: human trafficking, atrocities in the Bosnian war, counter-terrorism forces, extraordinary rendition. The freedom sought by Nico Bellic in America, the constraints on his actions in Liberty City, and the portrayal of the shadow government (not simply corrupt as in other GTAs) help define the implicit limits of the new war on terror. The third-person frame and representation of violence also serve a pedagogical or ideological function, extending the decorporealisation of the player and the character to distance violence from its political antecedents or visceral consequences. The desire for an authentic mapping of New York/Liberty City plays against the fantasy violence, but again the focus on the local scale of neighbourhoods seems particularly of interest. *GTA IV* must be analysed as a dominant cultural text, and in particular the configuration of new normals.

RESEARCH PROGRAM

In conclusion, I want to draw this parallel between the corporeality of the bodies and the claim to verisimilitude, particularly as it relates to the geopolitical imaginaries of these games. Both RTS and FPS games make some claim to the authenticity – in their geopolitical, technological, and technical frames, as we will see below. *Diplomacy* is based on a real pre–World War I map of Europe and realist interpretations of the balance of power and realpolitik. *Civilization* adds complexity, new technologies, and competitive, culturally-specific artificial intelligences to more closely approximate world history. *AA*, for example, talks about the modelling of their real weapons systems, uniforms, and tactics – and accords the success of the franchise to its realism and the participation of real heroes. *GTA IV* scanned and mapped buildings in New York City to provide a realistic simulacra for Liberty City. While each of these games invoke verisimilitude, the social and political actions that one is encouraged to engage in are completely unrealistic. Part of the pleasure within the large interactive worlds of *Civilization*, *GTA*, and *AA* is precisely in determining the limits of behaviour: can a player jump from a helicopter and survive? Can I use a nuclear weapon to blow up a rival city or a rocket launcher to blow up a fire engine? However, in these games, that freedom is radically constrained within tightly controlled metanarratives of the state and sovereignty, police and military control, and an imperial geopolitical imaginary. As Murray argues, "In this imagined space, the conventional social contract is suspended; however, this is not to say that what results is a lawless space. Rather games such as *Grand Theft Auto* represent rule-based, problem-solving environments that require creative solutions within a defined set of parameters."[95] It is these parameters which are politically interesting. Sid Meier, the creator of *Civilization*, among others says: "A game is a series of interesting choices."[96] In these games, the fundamental geopolitical rules of contemporary warfare are hidden by the appeal to authenticity – thus we see an emergent conservatism that renders the pressing ethical, moral, and technological questions of war or politics invisible, as the screen offers the illusion of choice. These war games, particularly those that claim to be free or nonlinear and authentic, make visible contemporary geographical imaginations.

And, in each of these games, the bodies of the dead and injured disappear, while the player's character endlessly respawns. Many games emphasise their physics and graphics engines, including 'rag-doll' effects for the behaviour of lifeless bodies. If the player's avatar is shown to die, it is often in cut scenes (in-game movies) or in animation after death. Kingsepp addresses the question of corporal and the disappearance of the dead:

> Death both is and is not the end: Your enemies pass away, but your own death is rather a temporary absence. . . . You die, but you either

resurrect at once or after a short while, or, in the worst case, after having re-entered the level, a process known as respawning. This is a privilege normally not extended to your enemies (when they die, that is it, and quite often, their extinction is accentuated by the total disappearance of their bodies).[97]

She argues, in particular, that there is an important sleight of hand: while the games claim authenticity, immersion, and historicity, there is a marked absence of death, what she terms "the paradoxical oscillation between the game's desire for representing reality (and the gamer's supposed desire for experiencing it) and their fundamental character of hyperreality. Still, when everything else has turned unreal, death would be the ultimate proof of being alive. But in the hyperreal, even death has lost its link to reality."[98] I want to draw a contrast between the actual or 'real' images of bodies – shown or censored in mainstream media – with the presentation of virtual bodies. Both 'self' and 'other' bodies are subject to differential corporealisation. The physical attributes of the masculine American body are abstracted from fitness, pain, injury, or emotions – even as those are the characteristics that define heroism or service. The attributes of the 'other' body are also abstract: nondescript in life and disappearing in death. The bodies of self are always regenerated and always whole; bodies of targets are always dismembered. This combination of realism and unreality has the effect of undermining the very project of representation: of undermining the representations of real conflicts, real dead bodies, real spaces.

Within the field of ludology there is much interesting research on the creation and maintenance of social networking in massively multiplayer online games (MMOs), the configuration of work in virtual worlds, narratological structures for sandbox games – all of which have important things to say for the geopolitics of video games.[99] This preliminary analysis of *Diplomacy*, *Civilization*, *America's Army* and *Grand Theft Auto* suggests three productive avenues for further study for critical geopolitics in the field of video games: popular culture and geopolitics, the question of scale in the representations of geopolitics and in particular the role of the city, and the role of the body in virtual worlds.

Engagement with the virtual + popular: One of the most fruitful areas for the critical geopolitics and International Relations has been the analysis of the everyday, the quotidian, and the popular.[100] An important part of this pop turn in critical scholarship has been a move away from "How does this artifact represent the world?" to a question of "What does this frame say about the possibility of representing the world?" Grayson et al. suggest that the project must examine "political subjectivities, the politics of affect and their constitutive narratives found in popular culture are not irrelevant to material processes such as production, environmental degradation, war-fighting or the pursuit of profit margins . . . it is also in the cultural imaginary that significant political battles are fought . . ."[101] While films and to a certain extent literature have been analysed, video games and other kinds of digital media are under-theorised and under-studied in critical geopolitics. Der Derian and Powers have examined the role of simulation and war-fighting; how can we push this forward to understand how soldiers and strategists themselves understand violence and war in a hyper-mediated field? Grusin has made an important contribution to this debate through the concept of "premediation," by which he means the social and political processes through which all possible futures are already contextualised and contained within a particular news-entertainment narrative before the event.[102] David Harvey, among others, have argued that it is easier to imagine the end of the world than it is to imagine the end of capitalism – in part because the end of the world has been premediated through apocalyptic representations as diverse as Gore's *Inconvenient Truth*, *Independence Day*, *Armageddon*, and the "*Left behind*" series.[103] What is missing is an analysis of the particular configurations of the political future in large video games franchises such as the *HALO*, *Bioshock*, *Fallout*, *Gears of War*, *Call of Duty* (including the latest iteration "Black Ops" which was released on September 11th 2010) and so on, which imply very imaginative premeditations of the geopolitical future. How are our understandings of the future (of politics and geopolitics) interpolated through video games?

Virtual/ urban: The city plays a crucial role in the imaginative geographies of these games, particularly *Civilization* and *GTA*. Although on different registers, the city represents in both games the seat of power, the centre of politics, and a particular spatial centre from which exclusive territorial control emanates. On the other hand, in *AA* the city is rendered as the complex space of particular war operations, filled with accessible and dangerous places (often driven through at speed or infiltrated and dominated). The scale of the tactical shooter is much smaller. However, whether as the setting for simple or complex problems of governance or control – each of these games could represent a supplement to the compelling analyses of military strategy by Graham and Shaw on urbicide.[104] The scale of urban conflict games as local/tactical and the demise of the strategic (hex) game view seem to demonstrate an interesting change in the geopolitical imaginary. So too could we examine the evolution of city-level games from *SimCity* (1989) through *States of Emergency* to *GTA IV*. One might even examine the evolution of the *Sim* franchise, from the control of municipal governance in the original series through ant farms in *SimAnt*, to *SimGolf* and *SimCopter* – but also the move towards the biopolitical in *The Sims* and *Spore*, the former emulates quotidian life in a Western country based on the individual's avatar, the latter emulates evolution from a single-celled organism through to the conquest of space and inter-galactic warfare (again, with large emphasis in both games on the scale of the city). There are also very interesting implications for the representations of crime and the urban in video games, as suggested by the preliminary analysis of *GTA IV* above and the literature on representations of race in the series. These tensions between the governance, biopolitical and tactical war-fighting representations of the city could be extremely productive for critical geopolitics.

Finally, much more research is needed from both the critical geopolitics and IR communities on the politics of the body and the microphysics of corporealism in virtual worlds. In addition to understanding the different conceptions of work and play that are so apposite for understanding the imaginative work that games do, we must also analyse how representations of virtual bodies interpolate our own understandings of the corporeal. I have suggested above that the differential corporealisation of self/other, and in particular the impossibility of death through the respawning of the self and the disappearance of the other, coupled with the appeal to verisimilitude, renders the body and in particular representations of pain impossible to understand in the virtual world. While the Abu Ghraib photos, in particular, have been examined and critiqued, there has not been enough sustained scholarly analysis of the pain of others in gamic space.[105] It has been my argument here that there is a profound connection between the appeal to realism in the representations of space and the body in the geographical imaginations of these virtual worlds, which have serious implications for our understanding of violence and empire.

ACKNOWLEDGEMENTS

The author would like to thank David Grondin for his work on this special issue, the editor and reviewers of *Geopolitics* who provided some extremely productive feedback. Aspects of this paper were presented at a number of conferences, including the 2009 World Politics and Popular Culture conference in Newcastle and much credit is due to the organisers and participants. My friend and colleague Debbie Lisle also read a very drafty version with great patience; Patrick Thaddeus Jackson has also been an important interlocutor on all matters professional, scientific, and fictional.

NOTES

1. While debates circulate in literary theory (between narratologists and ludologists) and new media studies over the terminology of "video" games, "digital" games, e-games, or "new media," I will use the first term. This can be seen as a continuation of Shapiro and Klein. Michael J. Shapiro, 'Representing World Politics: The Sport/War Intertext', in J. Der Derian and M. J. Shapiro (eds.), *International/Intertextual Relations: Postmodern readings of world politics* (New York: Lexington 1989) pp. 69–96; Bradley S. Klein, 'The Textual Strategies of the Military: Or Have You Read Any Good Defense Manuals Lately?', in J. Der Derian and M. J. Shapiro (eds.), *International/Intertextual Relations: Postmodern Readings of World Politics* (New York: Lexington 1989) pp. 97–112.

2. Simon Dalby, 'Warrior Geopolitics: Gladiator, Black Hawk Down and the Kingdom of Heaven', *Political Geography* 27/4 (2008) pp. 439–455; Klaus Dodds, 'Hollywood and Popular Geopolitics of the War on Terror', *Third World Quarterly* 29/8 (2009) pp. 1621–1637; Marcus Power and Andrew Crampton, 'Reel Geopolitics: Cinemato-graphing Political Space', *Geopolitics* 10/2 (2005) pp. 193–203; Michael J. Shapiro, *Violent Cartographies: Mapping Cultures of War* (Minneapolis: University of Minnesota Press 1997); Michael J. Shapiro, *Cinematic Geopolitics* (London: Routledge 2009); Cynthia Weber, *Imagining America at War: Morality, politics, and Film* (London: Routledge 2006).

3. Marcus Power, 'Digitized Virtuosity: Video War Games and Post-9/11 Cyber-Deterrence', *Security Dialogue* 38/2 (2007) pp. 273, 284.

4. Jason Dittmer, *Popular Culture, Geopolitics and Identity* (Lanham, MD: Rowman Littlefield 2010); Nick Dyer-Witheford and Greig de Peuter, *Games of Empire: Global Capitalism and Video Games* (Minneapolis: University of Minnesota Press 2010); Nina B. Huntemann and Matthew Thomas Payne, 'Introduction', in N. B. Huntemann and M. T. Payne (eds.), *Joystick Soldiers: The Politics of Play in Military Video Games* (London: Routledge 2010) pp. 1–18; Michael Longan, 'Playing with Landscape: Social Process and Spatial Forms in Video Games', *Aether: The Journal of Media Geography* 2 (2008) pp. 23–40; Power (note 3).

5. John Agnew, 'The Territorial Trap: The Geographical Assumptions of International Relations Theory', *Review of International Political Economy* 1/1 (1994) pp. 53–80.

6. The *Call of Duty, Half-Life* or *Halo* series, on the other hand, are 'rail' shooters, so called because even when presented with a choice of tactics, the game play is linear – the space of the game is likened to a rail, where players must progress down a particular, spatially defined track.

7. Chris Plante, 'All the World's a Sandbox', *Gamasutra: The Art and Business of Making Games* (12 May 2008), available at <http://www.gamasutra.com/php-bin/news_index.php?story=18545>.

8. Gonzalo Frasca, 'Sim Sin City: Some Thoughts about Grand Theft Auto 3', *Game Studies* 3/2 (2003), available at <http://gamestudies.org/0302/frasca/>.

9. Marc Augé, *Non-places: Introduction to an Anthropology of Supermodernity*, trans. by John Howe (London: Verso 1995).

10. Christopher Coker, *Waging War Without Warriors: The Changing Culture of Military Conflict* (Boulder: Lynne Rienner 2002).

11. James Der Derian, *Anti-diplomacy: Spies, Terror, Speed, and War* (Oxford: Blackwell 1992).

12. Elaine Scarry, *The Body in Pain: The Making and Unmaking of the World* (Oxford: OUP 1985) p. 62.

13. Shapiro, *Violent Cartographies* (note 2) p. 31.

14. Hartmut Gieselmann, 'Ordinary Gamers: The Vanishing Violence in War Games and its Influence on Male Games', *Eludamos: Journal for Computer Game Culture* 1/1 (2007) p. 4.

15. Coker (note 10) p. 173.

16. Gonzalo Frasca, 'Simulation versus narrative: Introduction to ludology', in M. P. Wolf and P. Perron, (eds.), *The Video Game Theory Reader* (New York: Routledge 2003) pp. 221–236.

17. Dittmer (note 4); Marc G. Doucet, 'Child's Play: The Political Imaginary of IR and Contemporary Popular Children's Films', *Global Society: Journal of Interdisciplinary International Relations* 19/3 (2005) pp. 289–306; Kyle Grayson, Matt Davies, and Simon Philpott, 'Pop Goes IR? Researching the Popular Culture-World Politics Continuum', *Politics* 29/3 (2009) pp. 155–163; Daniel H. Nexon and Iver B. Neumann (eds.), *Harry Potter and International Relations* (Ladham MD: Rowman and Littlefields 2006); Weber (note 2); Jutta Weldes, 'Popular Culture, Science Fiction and World Politics: Exploring Intertextual Relations', in J. Weldes (ed.), *To Seek Out New Worlds: Exploring Links Between Science Fiction and World Politics* (New York: Palgrave Macmillan 2003) pp. 1–27.

18. James Der Derian, *Virtuous War: Mapping the Military-Industrial-Media-Entertainment Network*, 2nd ed. (London: Routledge 2009) p. 26.

19. Entertainment Software Association, 'Industry Facts' (2010), available at <http://www.theesa.com/facts/index.asp>, accessed 18 June 2010.

20. Stephen Kline, Nick Dyer-Witheford, and Greig de Peuter, *Digital Play: The Interaction of Technology, Culture, and Marketing* (Montreal/Kingston: McGill-Queen's University Press 2003) p. 247.

21. See <http://www.fragdolls.com/index.php/gamer-girls>, accessed 13 June 2010.

22. David Leonard, '"Live in Your World, Play in Ours": Race, video games, and consuming the other', *SIMILE* 3/4 (2003).

23. Alexander R. Galloway, *Gaming: Essays on Algorithmic Culture* (Minneapolis: University of Minnesota Press 2006) p. 72.

24. Ibid., p. 92.

25. Aaron Hess, '"You Don't Play, You Volunteer": Narrative Public Memory Construction in Medal of Honor: Rising Sun', *Critical Studies in Media Communication* 24/4 (2007) pp. 339–356.

26. Weber (note 2) pp. 13–28.

27. Eva Kingsepp, 'Fighting Hyperreality With Hyperreality: History and Death in World War II Digital Games', *Games and Culture* 2/4 (2007) p. 368.

28. For this project, I am not considering the physicality or sociality of playing video games, although both may have value for further study.

29. Prof. K. J. Holsti introduced this game to myself, and generations of graduate students and colleagues, through an annual tournament, graciously hosted by himself. Richard Sharp, *The Game of Diplomacy* (1978), available at <http://www.diplomacy-archive.com/god.htm>.

30. Sebastian Deterding, 'Living Room Wars: Remediation, Board Games and the Early History of Video Wargaming', in N. B. Huntemann and M. T. Payne (eds.), *Joystick Soldiers: The Politics of Play in Military Video Games* (London: Routledge 2009) p. 23.

31. Allan Calhamer, 'The Invention of Diplomacy', *Games and Puzzles* 21 (1974), available at <http://www.diplomacy-archive.com/resources/calhamer/invention.htm>, accessed 12 June 2010.

32. Allan B. Calhamer, *Rules of the Game of Diplomacy* (1959), available at <http://www.diplomacy-archive.com/resources/rulebooks/1959.pdf>, accessed 12 June 2010.

33. Compare the rules of *Risk*, the nearest competitor to *Diplomacy*, in which the territories are sometimes but not always countries, and the retention or capture of a territory during a turn is rewarded with a card. Thus, the randomness of the cards, which lead to additional armies, puts chance at the heart of the game. Parker Brothers, *Risk: Rules of Play for Parker Brothers' Continental Game* (1959), available at <http://www.hasbro.com/common/instruct/Risk1959.PDF>, accessed 12 June 2010.

34. Duningan in Deterding (note 30) p. 26.

35. See Myers on the importance of *Civilization* in popular and gamer culture. According to its own press, the series has sold over six million units since its first iteration in 1991, and been listed as one of the top ten computer games by a number of sources over the past 19 years. David Myers, *The Nature of Computer Games: Play as Semiosis* (New York: Peter Lang 2003).

36. Galloway (note 23) p. 98.

37. Critics argue that the ideological goal of victory is inevitably an American-style liberal-democracy technocracy. Kacper Pobłocki, 'Becoming-State: The Bio-Cultural Imperialism of Sid Meier's *Civilization*', *Focaal: European Journal of Anthropology* 39 (2002) pp. 163–177; Tom Henthorne, 'Cyber-Utopias: The Politics and Ideology of Computer Games', *Studies in Popular Culture* 25/3 (2003).

38. Myers (note 35) p. 201.

39. R. B. J. Walker, *Inside/Outside: International Relations as Political Theory* (Cambridge: CUP 1993).

40. Ben DeVane and Kurt D. Squire, 'The Meaning of Race and Violence in Grand Theft Auto', *Games and Culture* 3/3–4 (2008) pp. 267–268.

41. Abhinava Kumar, *America's Army Game and the Production of War*, York Centre for International Security Studies Working Paper 27 (Toronto: York University 2004).

42. Galloway (note 23) p. 57.

43. Ibid., pp. 79–82.

44. Similar kind of analysis by Hansen of the Danish publication of the cartoons of the Prophet Mohammed. Lene Hansen, 'The Clash of Cartoons? The Clash of Civilizations? Visual Securitization and the Danish 2006 Cartoon Crisis', presented at the 48th Annual ISA Convention, Chicago, 28 Feb.–3 March 2007.

45. America's Army, 'Army Game Project' (2008), available at <http://www.americasarmy.com/about/>.

46. America's Army, 'America's Army Game Features' (2008), available at <http://www.americasarmy.com/intel/features.php>.

47. Randy Nichols, 'Target Acquired: America's Army and the Video Games Industry', in N. B. Huntemann and M. T. Payne (eds.), *Joystick Soldiers: The Politics of Play in Military Video Games* (London: Routledge 2010) p. 40.

48. America's Army, 'America's Army: Real Heroes' (2008), available at <http://www.americasarmy.com/realheroes/>.

49. Again, this is a topic of vigorous conversation on the *AA* forum. See: <http://forum.americasarmy.com/viewtopic.php?t=293423&highlight=women>, accessed 16 June 2010.

50. Kumar (note 41) pp. 6–8.

51. David B. Nieborg, 'Training Recruits and Conditioning Youth: The Soft Power of Military Games', in N. B. Huntemann and M. T. Payne (eds.), *Joystick Soldiers: The Politics of Play in Military Video Games* (London: Routledge 2010) p. 62.

52. Margaret Davis, Russell Shilling, Alex Mayberry, Phillip Bossant, Jesse McCree, Scott Dossett, Christian Buhl, et al., *Making America's Army, in America's Army PC Game: Vision and Realization* (San Francisco: US Army and the Moves Institute 2004) pp. 10–11, available at <http://www.movesinstitute. org/~zyda/pubs/YerbaBuenaAABooklet2004.pdf>.

53. Ian Bogost, 'The Rhetoric of Video Games', in K. Salem (ed.), *The Ecology of Games: Connecting Youth, Games and Learning* (Cambridge: MIT Press 2008) p. 130.

54. This is not always true in other games: The expansion to *Battlefield 2* permits players to choose "Insurgent" as their character. Nieborg, 'Training Recruits' (note 51) p. 54.

55. Nichols (note 47) p. 61.

56. Nina B. Huntemann, 'Interview with Colinal Casey Wardynski', in N. B. Huntemann and M. T. Payne (eds.), *Joystick Soldiers: The Politics of Play in Military Video Games* (London: Routledge 2010) p. 181.

57. America's Army, 'Mission: Border' (2008), available at <http://www.americasarmy.com/intel/ missions.php?id=36>.

58. David B. Nieborg, *America's Army: More than a Game Transforming Knowledge into Action Through Gaming and Simulation* (München: SAGSAGA 2004) p. 2.

59. One can see discussions of this at *America's Army* forum. For example: "The point is, its a joke. It is nothing serious about it, and it is a very fun way to spice up gameplay. I think AA staff should keep teabagging in the game. When your playing with friends, its more fun to teabag them when you kill them. It gives people a reason to fight- Teabag the fallen enemys. Go teabagging!" See 'Eliminate Teabagging' post (11 March 2008), available at <http://forum.americasarmy.com/viewtopic.php?t=274969>, accessed 16 June 2010.

60. Huntemann and Payne (note 4) p. 7.

61. Kumar (note 41); Power (note 3) p. 281.

62. Henry A. Giroux, 'War on Terror: The Militarising of Public Space and Culture in the United States', *Third Text* 18/4 (2004) p. 217.

63. Nieborg, *America's Army* (note 58) p. 6.

64. Gieselmann (note 14) p. 3.

65. James Der Derian, 'War as Game', *Brown Journal of World Affairs* 10/1 (2003) pp. 39–40.

66. Paul Barrett, 'White Thumbs, Black Bodies: Race, Violence and Neoliberal Fantasies in Grand Theft Auto: San Andreas', *The Review of Education, Pedagogy, and Cultural Studies* 28/1 (2006) p. 109.

67. Nichols (note 47) p. 41.

68. Giroux (note 62) p. 217.

69. Debbie Lisle, 'Comments at Cinematic Geopolitics Roundtable', 48th Annual International Studies Association, Chicago, 2007; Dyer-Witheford and de Peuter (note 4).

70. E. Casey Wardynski, 'Informing Popular Culture: The *America's Army* Game Concept', in M. David (ed.), *America's Army PC Game: Vision and Realization* (San Francisco: US Army and the Moves Institute 2004) p. 7, available at <http://www.movesinstitute.org/~zyda/pubs/YerbaBuenaAABooklet 2004.pdf>.

71. America's Army, 'Virtual Army Experience Fact Sheet' (2010), available at <http://vae. americasarmy.com/pdf/vae_factsheet.pdf>.

72. Ron Gould, 'The Virtual Army Experience: Mobile Event Case Study' (2008), available at <http://www.wholeinmyhead.com/Experiences_files/VAE.pdf>.

73. Will Freeman, 'The Virtual Army Experience', *news.cnet.com*, 1 Oct. 2007, available at <http:// news.cnet.com/8301-17938_105-9788521-1.html>.

74. Der Derian, *Virtuous War* (note 18) pp. 1–10.

75. America's Army, 'Recon' (2010), available at <http://www.americasarmy.com/intel/recon. php>, accessed 16 June 2010.

76. Galloway (note 23) p. 79.

77. Paul, Franklin, 'Take-Two's Grand Theft Auto 4 Sales Top $500 Million', *Reuters*, 8 May 2008, available at <http://www.reuters.com/article/hotStocksNews/idUSWNAS233520080507>.

78. DeVane and Squire (note 40) p. 264.

79. A similar game, released in 2002 by the same studio Rockstar Games, pushed this conceit even further: *State of Emergency*'s sandbox was a game world dominated by 'The Corporation' in which a rebel group 'Freedom' attempted to create anarchy. Within a mission structure that includes

political assassination and drug dealing, "the spirit of violent social upheaval seen in events like the Rodney King rebellion in Los Angeles [is transposed] into a participatory gaming environment." Galloway (note 23) p. 76.

80. This is a particular geopolitics of crime and punishment that maps well onto contemporary theorisation of "government through crime" and indeed the "war on terror." Jonathan Simon, 'Choosing our Wars, Transforming Governance: Cancer, Crime and Terror', in L. Amoore and M deGoede (eds.), *Risk and the War on Terror* (London: Routledge 2008).

81. Mikolaj Dymek and Thomas Lennerfors, 'Among Pasta-Loving Mafiosos, Drug-Selling Columbians, and Noodle-Eating Triads – Race, Humor and Interactive Ethics in Grand Theft Auto III', Proceedings of the Digital Games Research Association 2005 Conference: Changing Views Worlds in Play, 2005, p. 9.

82. Barrett (note 66) p. 95.

83. DeVane and Squire (note 40) p. 279.

84. Dymek and Lennerfors (note 81) p. 3.

85. Dean Chan, 'Playing with Race: The Ethics of Racialized Representations in E-Games', *International Review of Information Ethics* 4 (2005) p. 28.

86. Dyer-Witheford and de Peuter (note 4) pp. 156–158.

87. DeVane and Squire (note 40) p. 265.

88. Barrett (note 66) p. 101.

89. Dyer-Witheford and de Peuter (note 4) p. 171.

90. Stephen Graham, 'Imaging Urban Warfare: Urbanization and U.S. Military Technoscience', in D. Cowen and E. Gilbert (eds.), *War, Citizenship, Territory* (London: Routledge 2008) pp. 33–56.

91. Simon Bart, 'What if Baudrillard was a Gamer?', *Games and Culture* 2/4 (2007) pp. 355–357.

92. This was also true in the original *GTA* and *GTA: II*, which was shot in a top-down perspective (and still provides the perspective for the mini-map within the HUD).

93. Soraya Murray, 'High Art/Low Life: The Art of Playing Grand Theft Auto', *Performing Arts Journal* 80/27 (2005) p. 96.

94. DeVane and Squire (note 40) p. 272.

95. Murray (note 93) p. 97.

96. Stéphane Bura, 'Emotion Engineering: A Scientific Approach for Understanding Game Appeal', *Gamasutra: The Art and Business of Making Games* (29 July 2007), available at <http://www.gamasutra.com/view/feature/3738/emotion_engineering_a_scientific_.php?print=1, accessed 10 June 2010>.

97. Kingsepp (note 27) p. 371.

98. Ibid., p. 372.

99. Dyer-Witheford and de Peuter (note 4); Eric Hayot and Edward Wesp, 'Reading Game/Text: EverQuest, Alienation, and Digital Communities', *Postmodern Culture* 14/2 (2004); Greg Lastowka, 'Planes of Power: EverQuest as Text, Game and Community', *Game Studies* 9/1 (2009).

100. Francois Debrix, *Tabloid Terror: War, Culture and Geopolitics* (London: Routledge 2008); Dittmer (note 4); Derek Gregory and Alan Pred (eds.), *Violent Geographies: Fear, Terror, and Political Violence* (London: Routledge 2007).

101. Grayson, Davies, and Philpott (note 17) p. 157.

102. Richard Grusin, *Premediation: Affect and Mediality After 9/11* (Basingstoke: Palgrave Macmillan 2010).

103. Peter Y. Paik, *From Utopia to Apocalypse: Science Fiction and the Politics of Catastrophe* (Minneapolis: University of Minnesota Press 2010); Jason Dittmer and Tristan Sturm (eds.), *Mapping the End Times: American Evangelical Geopolitics and Apocalyptic Visions* (Aldershot: Ashgate 2010).

104. Stephen Graham, *Cities under Siege: The Military Urbanism* (London: Verso 2010); Martin Shaw, 'New Wars of the City: Relationships of 'Urbicide' and 'Genocide", in S. Graham (ed.), *Cities, War, and Terrorism: Towards an urban geopolitics* (Oxford: Wiley Blackwell 2004) pp. 141–153.

105. Judith Butler, 'Photography, War, Outrage', *PMLA* 120/3 (2005) pp. 822–827; Susan Sontag, *Regarding the Pain of Others* (New York: Picador 2004).

'Theatres of War': Visual Technologies and Identities in the Iraq Wars

BENJAMIN J. MULLER

Department of Political Science, King's University College at the University of Western Ontario, London, Ontario, Canada

JOHN H. W. MEASOR

Department of Political Science, Saint Mary's University, Halifax, Nova Scotia, Canada

The proliferation and development of media during the current war in Iraq and the opening of the Iraqi media landscape, which had heretofore been dampened through authoritarian state control and international isolation due to economic sanctions, has played a role in political articulation and identity formation throughout the occupation in a fashion not seen during previous conflicts. The adoption of media tools, emerging as both technology and its applications – often referred to as Web 2.0 – impacted both combatants and civilians caught within the conflict providing them the opportunity to capture and articulate their own experience in a fashion never before available. This investigation points to the potential impacts on current political action, the resourcefulness of opposition access to media, as well as dangers posed through the emotive content most often produced. We argue that while such media production is disassociated from surrounding events (or decontextualised), its nonetheless trusted and impactful due to its viral distribution and sourcing networks and diverse in its impact on those not intimately involved due to its lack of overarching narrative and production away from sites of power.

MEDIA IN THE 'FRONT LINE OF FREEDOM'

With the 2003 overthrow of the Ba'thist regime of Saddam Hussein, the end of both its state monopoly on information and authoritarian population controls, as well as thirteen years of isolation due to draconian economic sanctions, Iraqis joined the digital age. While the war and occupation ravaged much of Iraqi society this proliferation of media brought a previously dark and quarantined terrain[1] to the ever-expanding digital domain during a period of immense change within both domains. Both the technologies and the dynamic socially based uses of the 'wired world' invaded Iraq and Iraqi consciousness' alongside Anglo-American military forces. From an environment where the state controlled all media and altered much interpersonal communication – where all television, radio and print media were either controlled by the authoritarian regime in Baghdad or its opponents abroad – Iraqis found themselves in the midst of the digital revolution then overtaking the globe. The proliferation of the 'Arab sats', initiated by the 1996 launch of *Al-Jazeera* satellite news channel, as well as the establishment of widespread access to internet-enabled computers and telephones, afforded Iraqis increasing access to a freer flow of information as well as the ability to interact with the regional and global community.[2] With seemingly no restrictions remaining against free expression, celestial television and radio broadcasts, as well as print media, swamped the Iraqi marketplace. Political parties, militias, or anyone who could afford to broadcast a signal or distribute printed matter proffered information, entertainment, or social and political alternatives. Moreover, satellite dishes, mobile phones, internet cafes, and increasing www access allowed for an interactivity previously unfathomable under the Ba'th regime with its *mukabarat* (secret police), informants, and control over media that reached far beyond Iraq's borders.[3]

However, one should not assume that Iraqis were the only people within the territorial confines of the Iraqi state to be experiencing such an alteration of the possibilities afforded by the technological expansion of media-production to most members of society. Into this terrain marched the digital warriors of occupation forces, formerly exiled Iraqis, transnational social and political groups, and Islamist jihadists of many stripes.[4] The many competing social and political forces descending into the vacuum left in the wake of the decapitation of the Iraqi state were well familiar with the potential of the digital age and its political import. Whereas prior to the fall of the Ba'athist regime such forces had found themselves facing immense barriers to reach Iraqis, in the vacuum created by the decapitation of the state and the Iraqi people's desires to communicate with each other as well as the world previously denied to them, a rush of potentialities emerged. American soldiers, employees of the Coalition Provisional Authority (CPA), Nongovernmental Organizations (NGOs), Iraqi insurgents opposing the occupation, militias the country over and jihadist groups all needed to engage, promote and propagandise to Iraqis and the wider regional and global communities to raise support, justify their actions, and garner political legitimacy.

Within this mass, two groups quickly garnered our attention and led to this research project. The first were those guerillas, militants, jihadists, terrorists etc. who were opposing the occupation or US designs on the political formulation of Iraq's future. Clearly this represented a wide scope of the Iraqi political landscape and provided a window into the obvious diversity of the opposition movement. Much of their media creations were both overtly political in nature and corporate in production as they attempted to explain purpose, recruit support, and elicit a response. However, many of the video clips and web-based media creations also allowed for intimate and personal portrayals of the world views and expositions of participants in the insurgency and beyond. Often the terrain of competition became an evolving ecology of 'Iraqiness'. This opened our eyes to the obvious diversity found within – as well as the fictionalisation of – those opposing US occupation. These media productions undermined claims made by US leadership in Baghdad, MacDill AFB in Florida, Qatar and Washington regarding the 'Ba'thist dead ender' and 'al-Qaeda' basis' of those opposing the coalition's 'new Iraq'. Simply, insurgents and virtually all Iraqi-based groups were quick to produce media explaining their role and intentions, leaving little unsaid about their composition and visions for Iraq and its future.

However, a second group also emerged. US and coalition forces – both military and civilian – were creating large amounts of media. Like the Iraqis and foreign jihadists these soldiers and civilian contractors desired to 'get the truth out' and to speak with their own voices, primarily to those not in Iraq. Unlike the diverse target audience(s) of concern in the Iraqi case, US soldiers and Western contractors were almost singularly concerned with speaking to an audience "back home." Related to this is the fact that far more of the Iraqi-produced media had greater strategic significance in its construction, as it was often not simply expected to "tell a story", but rather speak to constituencies in Iraq and the broader Arab and Islamic worlds as well as express to their opponents their ability to create fear, mete out retribution, and to carry out sophisticated military tactics. However, US forces demonstrated more a fascination with war, the daily routine of hostilities and occupation duty, as well as attempts to lend granular reality to – and often in opposition to – the portrayal of their experience provided by an image-conscious and highly scripted Pentagon and media-savvy US administration. Overcoming not so much US government censorship – US forces and CPA personnel as well as contractors overwhelmingly supported 'the mission' – they were attempting to chronicle their own experience and overcome what they saw as a distorted mainstream media (MSM) portrayal of events in Iraq. This should not be interpreted as solely an anti-MSM perspective of the liberal-conservative political divide found in the United States. Rather, these contributions were predominantly attempts to humanise the experience and to record the reality of conflict and reconstruction – as these participants witnessed it – free of the antiseptic casualty-free gloss and filters of official

Washington and the MSM. Quickly, such creations spanned a gambit of intimate portrayals of daily routine or coalition member's attempts at reaching out to Iraqis in need against a background of often increasing violence and snuff film quality expressions of grotesque barbarity first from occupation tactics and then as Iraqi-on-Iraqi sectarian violence rapidly escalated. A particular exposure of such negative portrayals reached mainstream attention when a pornographic website based in the Netherlands began trading free access to pornography in return for US and coalition soldiers' videos and pictures of grotesque scenes from Iraq and Afghanistan – a clear enunciation of 'war pornography'.

Examining both streams of media has provided a window into the less varnished basis of the many political factions currently competing within Iraq for control over that territory, its peoples, and its immense natural resources. However, the 'battle in the wires' that increasingly allowed for a certain articulation of the battle of ideas and the creation of a great many potentialities for political and social action has now moved beyond the wires into lived experiences. These mediated experiences tell us much about the geopolitics of distance and affect, and contribute to spatialising war beyond the battlefield for those who consume these images. This has manifested itself in a great many expressions of 'blowback' as such media have emerged into news coverage of the conflict, taken on the mantle of 'being' the locus of contestation itself as much as the violence found across Iraq, and finally into MSM in the form of entertainment in both the Arab and Islamic worlds and especially in portrayals of the conflict within American prime-time television.[5] What we heretofore refer to as the 'battle in/for the wires', is along the lines of what James Der Derian refers to as the 'Age of Infoterror', wherein information itself becomes the critical 'force-multiplier of anxiety, fear, and hatred'.[6] In much the same way as Der Derian considers the back and forth between almost *ad nauseum* presentations of the collapsing World Trade Centre – sometimes to nefarious political end – and the grainy audio and video clips of bin Laden,[7] we consider the tussle for authenticity vis-à-vis mediated representations by both those claiming ownership over Iraqi identity and Iraqiness, and US military personnel making attempts to (re)tell the 'authentic' story of conflict.

The conditions of possibility for much of this lies in the extent to which – particularly in contemporary Western societies – warfare is primarily a mediated phenomenon. In this regard, the battle in the wires – or salvo in the 'Age of Infoterror' – is to a great extent *the* battle. Through meditations on Heidegger's warning that we ought not try and understand technology technologically, Arthur Kroker proposes "uncanny thinking," which embraces and responds to Heidegger's contention that it is not the technologisation of the globe that is uncanny, but rather our total lack of preparation for this transformation.[8] There appears a similar dilemma here. Rather than simply regarding these digital narratives as cogent and powerful representations of the battlefield, competing identities, and national and sub-national performances, but moreover to be conceived of as potentialities for political and social action. Thereby the analysis takes us into what Kroker refers to as, "a way of 'dwelling' between the irreconcilabilities of the fully objectified beings of technological society and as a yet unappreciated, and certainly until now silenced, way of *being technology*."[9] As Der Derian suggests in more direct reference to the sort of puzzle we find ourselves grappling with here: 'Global media plus global terror has tarnished the early promise of the Information Age. To get back to the future, we need to have a better picture of how media contributed to the realignments of power and knowledge that followed 9/11'.[10] Although we do not propose the extensive exegesis of pre- and post-9/11 media as Der Derian effectively forwards throughout his work,[11] our analysis grapples with a similar dilemma of the relationship between media actions, mediated actions, and actions of media. Again, as Der Derian claims, it is a puzzle related to the 'oscillations of message and medium (signal-to-noise ratio), [and] regressive repetitions of images (feed-back loops)'.[12]

SPECTACLES OF IDENTITY AND/IN CONFLICT

The insurgency in opposition to the project of a 'new Iraq' imported on the back of Anglo-American military conquest, has continuously morphed over the years of occupation. Responding to the new potentialities presented by the destruction of the Iraqi state 'e-insurgents' quickly adopted new technologies that had been increasing in popularity the world over. Retroactively referred to as 'Web 2.0' its development tends to be cited for the sudden rise in 'user generated content' such as YouTube and related manifestations of social networking. It is well situated within an Appadurai-like 'mediascape', where the ties between the real and the fictional are actively blurred both by medium and message.[13] As 'early adopters' of such technologies, Iraqis were often doing so alongside and at the same time as the American and coalition forces sharing the physical Iraqi landscape with them. Both groups quickly outpaced, and have remained largely quite far ahead of, official attempts to utilise such technologies and thereby compete with the visions they have created. This is unsurprising considering both Appadurai's own accounts of the proclivity of mediascapes to, as Michael Dartnell suggests, 'serve as important bases from which to fabricate ways of looking and doing'.[14] However, initial attempts to express a counter to the official US-led coalition's vision for Iraq should be examined to better understand the extent and viability of these various oppositions.

Initial forays began with the conflict itself rather than sophisticated political formulations. This is unsurprising as producing representations of battle, warriors, battlefields, and the like, is an integral part of violent political conflict itself.[15] Whether on the pottery or frescos of the ancient world, the daguerreotype photos from the Mexican-American war in the mid-nineteenth century, or the prolific war films produced during and shortly after World War II, representations of war and its participants have always had an integral role to relay the spectacle of violence to the wider citizenry. Thucydides' account of the Peloponnesian War, Herodotus's histories, the lionisation of 'Sgt. York' or Audie Murphy as much as Brent Sadler's appraisals on CNN of the hastily abandoned military hardware of Iraqi forces in Tikrit, all allow these observers of violent conflict to make such events legible for the wider citizenry who are not directly present.

Such accounts and representations of war are often translated into foundational mythologies, national narratives and cultural performances. The battles of the Peloponnesian War become paradigmatic of Greek city-state identity; the widely broadcast images of the so-called 'highway of death' from the 1991 Gulf War quickly became symbolic of Western military superiority;[16] and bombarded Serbian police installations during NATO operations in Kosovo served as testament to the use of precision munitions for humanitarian purposes. In a related manner, footage of body bags returning from Vietnam and images of the Mei Lai massacre; the sight of Afghan children sitting among unexploded cluster bombs and food packages – whose colouring and size made them relatively indistinguishable from one another – and, tortured prisoners at Abu Ghraib and Guantanamo Bay, confirm the suspicions of some and raise the ire of others, precipitating intense critical reflection, ambivalences, and counter-performances of nationhood.[17] Whether expressing some guttural denial of identity and subsequent counter-performance of nationhood, or heartfelt patriotism and pride, the spectrum of responses are not often premised on direct relations to participants in conflict or physical presence in war zones, but vis-à-vis the representations of conflict; again, the oscillation between medium and message – or what Der Derian refers to as 'signal to noise ratio',[18] is essential.

What is the significance of representations that are produced not by observers but by participants? Raw articulations of those participating in the conflict now have the ability to capture their version of events, often with specific focus on the most brutalising impacts of human conflict. These spectacles, their production and wider role in the politics of conflict and identity form the focal point of our consideration here. In an age where the contemporary nation-state is countering numerous challenges to its political, geographic and economic prerogatives a diverse set of voices are emerging to re-articulate 'the nation'. While non-state actors have long competed to articulate such messages, micro- and informational technology advances now allow for small groups and individuals to prepare artistic productions of a sophisticated nature outside state cultural governance.

Our initial analysis concentrated on visual material produced by the combatants themselves: Whether self-made video clips by US forces in Iraq that depict gruesome events in the theatre of war, or spectacles that simply fetishise American military technology, as well as 'attack' videos made by forces opposing the US-led military occupation. In sharp contrast to the 'embedded' journalist whose ubiquitous military chaperon makes even the most trusting viewer highly sceptical of the 'objectivity' of the reporting, these often brutal but nonetheless novel visual spectacles making war possible/visible beyond the battlefield capture conflict in a raw and heretofore inaccessible fashion for those not directly participating in the conflict.

What then is the impact of these often graphic and grisly visual productions on the articulation of self and other in theatres of war? Do these visual spectacles become a part of the violent performances of war itself? The intended audience of such 'clips' and videos is varied and many productions clearly situate differing messages for differing constituencies.[19] Demonstration of military prowess and technical capabilities, evidence of honour in the practice of combat, demonstrations of the privations of life under the harsh conditions of modern warfare, and the 'othering' of opponents, all speak with a particular claim to 'authenticity of experience' that in its grainy particularity cuts through the gloss of sanitised and neatly packaged professional media. That insurgent videos in particular increasingly became the sole window on the primary events in Iraq – where news media could not operate freely due to violence and their increasing identification as participants on one particular 'side' to the conflict itself – therefore speaks to new potentialities for such actors to articulate the nation outside state- and corporate-controlled media across the globe. In contrast to Western media, who have little by way of hard copy to offer viewers, the enduring images of the current conflict will clearly emanate from the lenses of its participants – perhaps for the first time in history.[20]

We begin our analysis with some discussion of the physical material to which we are referring. In part, this is a burgeoning collection of videos ranging from slide-show images of US military hardware accompanied by rock music, to 'attack' videos made by oppositional forces which expose the vulnerability of seemingly invincible US technology and tactics, and finally to gruesome footage of battlefield deaths, execution-style killings, and maimed bodies. The theoretical considerations about the impact of such material, its possible role in constructing Self/other in theatres of war, the attempt within to forward particular articulations of identity, whether that be Arab, Islamic, or American, and the impact such visceral material has on what are supposedly non-participants or non-combatants in war provided an immense ecology of representations as these 'productions' proliferated.

Like the shared digital photos of prisoner abuse at Abu Ghraib the exposure of an exchange of photographs and video recordings from the conflict zones of Afghanistan and Iraq in return for amateur pornography on a Netherlands-registered and Florida-based website first brought such footage to the attention of the broader world.[21] The offer promoted by www.nowthatsfuckedup.com: 'If you are a US soldier deployed to Iraq or Afghanistan or any other theatre of war and you would like free access to the site, upload the photos which you and your buddies took during your service.' The most infamous photo, soon to proliferate across the Internet and bring attention to the site, was that of a human face, as though surgically removed from the rest of the head, found at the scene of a bombing in Iraq. Two shots, one of it on the ground and a second of it inside a plastic serving bowl were presented with commentary calling into question the humanity of Iraqis. The gory nature of the materials brought attention and criticism, as well as investigation by the US government. The taking and posting of such 'trophy' photos of dead Iraqis was a potential violation of the Geneva Conventions and damaging politically on the heels of the exposure of the torture of Iraqis by US service personnel at Abu Ghraib prison.

That US service members were creating such images – and propagating them to wider audiences via the internet – largely using personal digital technology, presented a new forum for participants of conflict to present personal accounts of 'their' war to make room for other, less familiar, spaces of war. The subsequent release of graphic photographs of the bodies of Uday and Qusay Hussein, the capture and medical examination of Saddam Hussein, and then the body of Abu Musab al-Zarqawi by US authorities did little to dissuade individual service members from such practices. This pornographisation and fetishisation of 'trophy' images and the dissemination of emotional stimuli was the extreme edge of 'othering' Iraqis for some American soldiers and their chosen audience. This process requires not only grotesque images to confirm such socialisation. In one pointed video, unremarkable in its presentation of a standard night-time US patrol, soldiers are seen repeatedly playing with an adopted Iraqi cat – now the unit's mascot. In playful guise and in joking as the cat accompanies them on their nightly rounds such scenes may prove unexceptional except in the moniker chosen – 'hajji cat'.[22] That such a cat could not be seen as simply a cat, but had to be 'othered' due to its location and nominal national origins spoke to the soldiers' larger process of 'othering' all things Iraqi from their wider norms. A cat is not a cat when its in Iraq . . . it's a hajji cat.

Now exposed to a wider audience, an examination of the ever-increasing catalogue of available electronic representations emanating from US military personnel serving in Iraq evidences much less graphic imagery. Much focuses on the mundane privations of life in conflict zones living under military regulations and unhappiness at extended tours or the decisions of senior officials (especially in the provision of equipment). Most prevalent, however, are demonstrations of the pride in service and good acts performed within both theatres of operation (Iraq and Afghanistan). Smiling Iraqi civilians, playful children, and successful reconstruction projects portray a positive image of the experience often missing from news media coverage that instead focuses on the spectacular and immediate political actions of the day such as car bombings, suicide bombings and the increasing sectarian-based deployment of death squads. Imagery otherwise, and predominantly, focuses on the established narratives of American political culture: pride in service of country, patriotism, the moral superiority of such service, and in the slavish appeal to technological and operational superiority and over-whelming firepower. Set to rock music and using images gleaned from service, as well as lifted from military sites across the internet, such videos glorify service and those who serve, often in classic cinematic formulations from Hollywood war movies. Indeed, 'mash-ups' of Iraq footage, favourite popular music and Hollywood scenes from movies, television and video games increasingly appear. The melding of long-standing patriotic imagery, the basis of US military esprit de corps, and popular imagery created by Hollywood of what it is to 'be' American shows an acceptance of many popular enunciations.

Unwilling to limit himself to a mash-up, aspiring singer Joshua Belile, serving as a corporal in the US Marines and deployed to Iraq, instead wrote and performed an increasingly popular song amongst those he was serving with that gained fame when distributed online over YouTube. Belile's performance of "Hajji Girl" initially garnered censure because of his depiction of the violence of a house search by US forces and his use of the younger sister of an Iraqi girl he had fallen in love with as a human shield when Iraqi insurgents attack his unit. While the broader public expressed reservations about the violence, and American Muslims protested his depictions of Islam and Islamic people, Belile – following a military investigation – faced no censure as his art failed to breach any military rules regarding the use of media equipment or conduct in the field.

In the age of so-called 'reality television,' one can also see an increasingly blurred distinction between the polished Hollywood imagery and the soldier-cum-film-maker's representations, exemplified in the recent documentary "Combat Diary: The Marines of Lima Company", screened on the A&E network in North America.[23] Interspersed with post-conflict interviews of the surviving members of an Ohio-based Marine unit, 'Lima Company,' and typical American scenes of returned Marines shopping with fiancées in suburban American strip malls, "Combat Diary" relies heavily on footage shot by the Marines themselves. Losing 23 of its 184 members during their seven-month tour of duty, Lima Company appears a logical focal point for the brutal costs of war, and as reviewers have noted, reliance on footage of their own 'puts a human face on those slain.'[24] As Brian Lowry indicates: 'We see them kicking in doors, exploding mines and improvised explosives and engaging in firefights, all the while creating a video archive – processing the war as their own real time movie.'[25] Although much of the material we are labelling as 'war pornography' has a rather limited audience, often targeted at fellow combatants, documentaries such as "Combat Diary" gain a much wider audience for these self-representations of combat and other service. While generally far more sanitised than the more gruesome and grisly imagery, these articulations of conflict nonetheless present a 'face of war' as seen through the eyes of combatants to the general public who are for the most part non-combatants. Moreover, presented as a 'real' face of war, such representations tend to emphasise the esprit de corps of US forces, and in the case of "Combat Diary", unique trust and reverence for Iraqi Special Forces.[26] Interestingly, while countering the typical 'othering' strategies of many pieces of 'war pornography' considered in this analysis, the introduction of Iraqi Special Forces to the unit is initially accompanied by hostility, mistrust, and scepticism. Although suspicions are overcome by the Marines in Lima Company, the interviews that form an integral portion of the documentary indicate how these initial misgivings were the result of racial and cultural stereotypes, and what in academic language would clearly fit the 'Orientalist' imagination.[27]

According to Michael J. Shapiro, cinematic representations have always had an integral role to play in constructing 'faces of war.'[28] One face is what might be termed the 'official' face, articulating the well-rehearsed trope that suggests we live in a dangerous world that must be confronted with force.[29] The other face, according to Shapiro, is ontological rather than strategic; focused on performing/asserting identity rather than the effects and 'necessity' of the use of force.[30] Such 'faces of war' are most obviously palpable in the popular imagery à la Hollywood, depicting celebrated forms of militarism and esprit de corps. However, while these recent participant-produced representations are frequently grotesque and disturbing, there are ways in which they subconsciously emphasise the 'official' face of war. Whether the grisly images of combat casualties, or even the disturbing fetishisation or pornographisation of 'trophy' images, like those mentioned earlier, such 'faces of war' tend to underscore in almost Hobbesian fashion the menacing prospect of statelessness. Rather than raise questions regarding the propensity towards brutality in conflict, and the loss of our common humanity, such 'nasty and brutish' displays, to borrow from Hobbes's well-known dictum, can be articulated as further proof that we live in a dangerous world that must be confronted with force. As the now infamous photographs from Abu Ghraib demonstrated, such disturbing imagery might not lead to anti-war sentiment, or even critical reflection on the necessity and possible consequences of military action, but simply as an aberration to the rule and the business of war as usual. Private contractors have run amok; command structures have failed; the Hobbesian 'daddy-state' was not sufficiently in control. In other words, such 'faces of war' can be cast as a reminder of the menacing 'state of nature' so exploited by the social contract theorists of modern political theory, rather than a visual representation of the damaging physical, cultural and psychological impacts of warfare.

These videos and representations have never stopped emerging from Iraq and Afghanistan. Powered by the imagination of the participants in the conflict and the accessibility of the internet to expose ever-increasing numbers to their existence, what changed, both within the experience itself, the focus of much of their output, and within the broader global wired society as well was the introduction over the 2003–2005 period of a series of technologies allowing for increased interactiveness – often along established social networks – known as 'Web 2.0'.

WEB 2.0 AND 'BLACK GLOBALISATION'

Iraqis quickly managed to connect with each other and the broader global community through mobile phones, internet cafes, and even home-based computers utilising Web 2.0 tools[31] such as blogs and social-networking sites, as well as picture and video-sharing sites such as Flikr and YouTube. Joining the digital age at the peak of the 'Web 2.0' explosion of user-generated content, while being occupied by forces many of whom were committed to the use of such technologies, as well as globally attuned insurgents steeped in the use of such technologies for communications, propaganda and operational efficacy – those Iraqis capable of accessing the online world were afforded the ability to articulate their own experiences rather than passively have foreigners craft their story. The ease with which 'form' and 'content' were separated across the digital domain by Web 2.0 software makes possible Kroker's "dwelling" between the objectified beings of technology and ways of "being" technology. To the bemusement of Heidegger, rather than focus on our lack of preparedness for this transformation and our inability to confront its potentialities, we remain preoccupied with the phenomena of the technical world itself. In addition to the very typical discussions that speak of globalisation in terms of flat worlds and global villages, articulations of technology as separate from humanity and neutral are ubiquitous. In the proliferation of models of the biometric "risk" state, beholden to particular technological channels towards security vis-à-vis surveillance, biometrics, and the subsequent (re)conceptualisation of citizens, migrants, refugees, and the like as "digital bodies," the dialogue is framed by the suggestion that technology is wholly separated from good or bad uses. Contrary then to the typical prosaic accounts of "black globalisation" as referring to phenomena such as terrorist (ab)uses of transportation and communication technologies and illicit trade flows, our analysis has something else in mind.[32] As Der Derian asserts in response to the sort of linear and progressivist teleologies of information technology forwarded by Thomas Friedman and others, it is increasingly difficult to discern the directions and developments of media as defined by an all powerful actor.[33]

Digital Warriors in the Battle of/for the Wires

Increasingly international media coverage of the conflicts from across the Muslim world have exposed the existence of videos showing speeches by jihadist Islamic militants such as Osama bin Laden, 'martyr' videos made by suicide bombers in Palestine or Chechnya, and 'beheading' videos made of the grisly murders of individuals captured by extremist groups. However, examples are no longer limited to the Middle East or Islamic world. Mexican criminal gangs began employing YouTube to post videos of the torture and execution of members of opposing criminal organisations in the mid-2000s. Demonstrations of such power are meant to intimidate opponents, dampen potential law enforcement, and terrorise the broader society. Where prior to the emergence of 'Web 2.0' technologies the censorship of audience sensibilities and political fears by MSM broadcasters allowed such media to remain obscured, sensationalised, and left unexamined outside the kiosks of the markets of the Islamic world or the obscurity of the internet. However, hundreds if not thousands of such videos now proliferate, available for illicit sale across three continents in street-front shops and *souqs* from Cairo to Damascus and Islamabad to Riyadh. Moreover, they are increasingly aired as both background material and news items on Arab and international news media, vastly increasing the bandwidth of those previously limited to internet forums. That they are recognised as a principle weapon of the Islamist movement is evidenced by the designation of specific 'media sections' from the many diverse elements and groups of the global jihadist movements. However, they are not limited to Islamists, nor highly organised political groups. Egyptian human rights organisations brought new life to campaigns to end ubiquitous police violence against citizens as well as misogynist street debasement of women through the use of video captured by mobile phone cameras evidencing such behaviour. The screams of prisoner's being tortured by Egyptian police officers led to the prosecution of such officers – and to the hypocrisy of the Mubarak regime being further exposed as it prosecuted the victim of the torture as well as those disseminating the clip on the Internet in addition to the police who carried out the abuses.

Ranging from half-minute video 'clips' to feature-length movies the videos produced by Iraqi insurgents and militias increasingly evidenced sophisticated production quality and layered messages emanating from an expanding number of groups in Iraq and beyond. The vast majority are unvarnished and often lacking in any commentary simply exhibiting images of the engagement of US military forces across Iraq. Obviously these rudimentary 'clips' have become de rigueur on internet sites run by the insurgents themselves as evidence of their operations against US forces, as a recruiting tool to excite those wishing to volunteer to fight, and as a fundraising tool to both gain attention with and prove the efficacy of support to those capable of donating funds. Clips showing the use of snipers against US sentries and patrolling soldiers, ambushes of US patrols demonstrating the capacity to engage larger and previously seemingly invincible forces, the destruction of supply convoys and the explosions of HUMVEEs by roadside IEDs (improvised explosive devices) were updated on an ongoing basis to further demonstrate ongoing operations. Many of the videos displayed military training with mortars and small unit tactics as well as the capacity to control the terrain of operations with videos showing the aftermath of hostilities – as the filming continues uninterrupted with the arrival of American, Iraqi, and coalition reinforcements to support and then clean up following the attack. However, the impact of such clips is limited largely to the narrow interests and audiences outlined above.

While such clips demonstrate operational efficacy and the broad geographic scope of anti-US militancy, they do not speak to a wider audience. Increasingly videos of a longer duration have emerged that attempt to do much more in pieces set to Arab and Islamic music and poetry in attempts to re-narrate the nation and call for support of jihad in Iraq and beyond. Often eight-to-ten minutes long they combine still photos and video footage to the sounds of well-known Arabic and Islamic poetry and nationalist music. Images of combat with US forces are interspersed with imagery of Iraqi historic sites and classic imagery of past glory, military engagements, and even unit insignia and monikers from the eighty-year history of the Iraqi Army. Moreover, US forces are shown in compromising imagery virtually always pulled from American and European online news sites. Visual evidence of US forces arresting, abusing and torturing Iraqi civilians playing on Arab, Islamic, and broader humanitarian senses of wronged nobility and honour undergird more focused political advocacy. Photographs of US soldiers crying are prominently juxtaposed with stern and valiant fighters of the Iraqi resistance. Coming in for repeated attention is the US attacks on Fallujah in April 2003 and November 2004, which has taken on in defeat the status of legend. Fallujah's ferocity of opposition to US incursions and eventual occupation as well as the immense destruction and commensurate humanitarian devastation of the city led many to suspect atrocities were committed by US forces.[34]

The videos, while utilising an idiom and cultural frames of reference of a broadly Islamic nature – thereby often indecipherable to a non-Arab speaking international constituency – also embed deeply symbolic and highly emotive commentary in Arabic. The competition of messages, ranging from Iraqi nationalism to al-Qaeda-inspired 'clash of civilisations' covers a wide range of groups fighting against US forces displaying a variety of imaginaries. Competing for adherents and political legitimacy they utilise images of holy sites, the Iraqi national anthem, the Shaheed Monument (commemorating Iraqi soldiers killed in the Iran-Iraq war) in Baghdad,[35] and ubiquitous shots of sandal-wearing resistance fighters as they engage body-armored and helmeted US forces in urban warfare. Many such examples demonstrate a re-appropriation of such images from their Ba'thist past.[36] Such capacities – to utilise previous images and motifs at relatively little cost using pre-existing online resources – allow for a repackaging of state-inspired imagery of the past in new guise(s). Moreover, while Anglo-American occupation forces are ubiquitously at the centre of such opposition, insurgent and other political groupings across the Iraqi spectrum share in their tying such forces to the nefarious intentions of regional opponents. Thus, while some Iraqis see the hidden hand of Iranian machinations in guiding American actions, others see that of Wahhabist or Salafi Islamists from the Arabian peninsula, and even those of Zionism as being ultimately responsible for the savagery they have been inflicted by and to explain the barbarity exhibited within their national space. Such Iraq-centric imagery is, however, widely tied to larger themes from across the Arab swath of the Islamic world. In a video by the 'Army of the Mujahidin in Iraq', the narrator cries, "We didn't only lose al-Aqsa,[37] they also then stole the K'aba.[38] Attack! Attack! Attack!" The use of oppositional Islamist poetry from Palestine to describe the struggle against occupation in Iraq and the repeated references to American infidel occupation of Mecca following the 1990–1991 Gulf Conflict merge into a global narrative of opposition to Western occupation of the Arab and Islamic world.

Not so dissimilar from the special issue of *Millennium* focused on the 'Sublime', our own analysis loosely fits into what some have termed the 'aesthetic turn' in international relations. Although the general focus is on the imagery of warfare in this case, the interest is not directed at how official films, news media, etc., produce depoliticised and amoral images of warfare, or indeed not so much regarding the impact of the mediated nature of warfare today on contemporary societies. These are important issues, integral to the so-called aesthetic turn in IR, and worth serious consideration.[39] The focus here is more with the need to rethink the battlefield, identity, and being political – something the user-generated content provided by combatants makes evermore salient to our understandings, and certainly the fecund space of Appadurai's 'mediascapes', mentioned earlier, in which the line between the real and the fictional is blurred to often great political affect. While these visual representations of conflict have the potential to screen some of the cultural, political and social implications of war, the analysis here is concerned with the ramifications of not simply representations of the battlefield, but the transformation of digital warriors in the battle of the wires. When insurgents carry out attacks like Hollywood producers, at times driven by the need to record and transmit the visual spectacle as much as they are by operational considerations, the battlefield and its participants require a deep rethink. Attacks on US forces are now filmed, often from multiple camera angles, and with high-resolution cameras. The footage is slickly edited into dramatic narratives: quick-cut images of Humvees exploding or US soldiers being felled by snipers are set to inspiring religious soundtracks or chanting, which lends them a triumphal feel.

WAR PORN AND THE POLITICAL

In one of his more recent texts, *The Politics of Aesthetics*, Jacques Ranciere has a rather intriguing account of politics that is worth noting:

> Politics exists when the figure of a specific subject is constituted, a super-numerary subject in relation to the calculated number of groups, places, and functions in a society.[40]

Ranciere underscores his understanding of politics as a struggle for an unrecognised subject or party to gain equal recognition within the established order. In *The Politics of Aesthetics*, what is deemed permissible to say or show in society is the focus for Ranciere, but it is also obviously connected to politics, which is for Ranciere captured in the complicated relationship between the aesthetics of politics and the politics of aesthetics. As elsewhere in Ranciere's work, the importance of how the subject understands itself in relation to others is vital.[41] In the case of these visual spectacles referred to here as 'war porn', there is clearly a concerted attempt to not only depict a particular account of 'the real', but also assert a self-understanding of the subject. As an attempted subversion and counter-performance to the mediated and cinematic representations of the subject, these visualisations are an essential part of Ranciere's struggle for equal recognition of the subject. However, for our project, of arguably greater importance is the extent to which the medium is also the message, insofar as the insurgent's ability to tell their own story vis-à-vis technology is a distinct part of their assertion of their subjectivity and the story.

In addition to Ranciere's political ground over which the permissible and impermissible is contested, these artifacts of 'war porn' fit well within Appadurai's notion of the 'mediascape'. The sorts of issues Ranciere raises regarding attempts to both articulate 'the real' as well as a self-understanding of the subject is for Appadurai, framed in terms of the large and complex repertoires of images, narratives and ethnoscapes provided, that sit over and above technological capacity.[42] Here is a long-standing connection with contemporary YouTube postings from US military personnel, or insurgency videos rapidly transferred across cellular networks, and the representations of the US Civil War, Ghandi's resistance to colonial oppression, and so on.

Blowback and Public Diplomacy

As the proliferation and the volume of the media increased, its production quality improved, and its content became more reflective of expositions of such warrants of recognition. Partly a matter of technological adaptation, partly a consequence of more organised content providers (especially with Iraqi political movements replacing many smaller operations), such alterations only underscored the media clips' increasingly salient and powerful role in political expression and engagement over the conflict. While the productions began with a level of intimacy eliciting a sense of 'the real', production quality and content was often reminiscent of 'home' video attributes. Americans and Iraqis alike suffused the mundane with performance of nation. Clips evidenced the particularity and diversity of those engaged in both the savagery of the conflict and the mundanity of occupation. Gradually such messaging altered as combatants – through 'media units' amongst Iraqi political factions and the free-and-seemingly unlimited usage of media tolerated by the US chain of command – saw creators of content increasingly desirous of clarifying their message and communicating directly with their presumed audience(s).

Who was such media eliciting support from? Americans were focused on garnering the support of the US public, with special focus on family and friends 'back home'. While our focus here has been on media creation by those inside Iraq, their access to the real-time debates and coverage of events being presented by mainstream American media clearly elicited responses. This was especially the case when equipment proved wanting in the harsh climatic conditions and dynamic combat brought about through rapid improvisational asymmetrical warfare tactics used by insurgents, when deployments were extended beyond contractual and promised limits, when political commentary in the United States impacted events and levels of violence in Iraq itself, and when official commentary to assuage family members – concerned over VA medical care, increasing levels of violence in Iraq, and an ever-shifting justification for extended deployment schedules – was made apparent to be counter to the experience of volunteer soldiers serving in Iraq.

While civilian control over the military and chain-of-command discipline both blunted even a hint of insurrection – the immediacy of reactions by serving members of the military will need be a concern for politicians and political commentators alike moving forward. Bravado such as President Bush's 'bring it on', arguing as Donald Rumsfeld that you fight 'with the army you have, not the army you want', and Tom Tancredo's call to 'take out' Mecca in the face of another attack on the United States may have carried domestic political appeal – but raised differing reactions amongst US service members then in Iraq. Talking heads on news analysis shows, policy wonks espousing solutions to assumed concerns, and politicians' general grammatical assumption to be speaking on behalf of soldiers and their interests may not have been a primary focus of many enlisted men – indeed their 'work' kept them far too busy – but the general conclusions and rhetorical embellishments enunciated 'back home' periodically solidify into calumnies to those 'over there'. Speaking over the anodyne productions of US military media-relations on the one hand and the overtly and incendiary political on the other US service members increasingly attempted to clarify for friends and family what their sense of mission was in their own words.

While in general agreement with American exceptionalism, the proscribed mission as assigned, and the general virtue of American efforts in Iraq, corrections to commonly assumed notions informed many such videos.[43] Common refrains included the dramatically higher levels of violence occurring both against US forces and between Iraqi factions than was assumed by media presentations and political commentary in the United States, the failure of Iraqi colleagues to be consistently professional and determined in their efforts to bring security to all Iraqis, and the enormity of the task involved in rebuilding a devastated Iraqi infrastructure and a seemingly absent political will for comity by Iraqi elites. Such notions were not sounded in defeatist terms, nor in opposition to tactical and strategic considerations evidenced by the chain-of-command, but rather in response to media, political, and familial portrayals to which they were privy. At no point was this more evident than during the 'surge' operations of 2007. Media created by service members directed to family evidenced a more nuanced view of the operations and sophisticated explanations for the decline in violence often encapsulated in a tone concerned for what would follow either complete American withdrawal or draw-downs of US forces. Such clips were in contrast – and often in response – to the mainstream media and political punditry's overtly positive portrayal bordering on a narration of 'victory' that US soldiers were not eliciting in their own productions.[44] One might question if the Bush administration's great "success" in Iraq has been in keeping so much of these soldiers media messages outside mainstream consideration.

Unlike the experience of professional American media outlets – their Arab counterparts and many European media providers openly embraced the production of such user- and participant-created content early on. This saw more professionally produced Iraqi videos as they increasingly framed their message to a broader audience, heightened political commentary at the expense of eliciting intimate support, raised their struggle to higher levels by tying it to occupations in Palestine and Chechnya and moreover in opposition to American global hegemony. Most noticeably videos increasingly became terrain for contested notions of Iraqi identity – easily differentiated by whom the creator of the media portrayed as 'other' or the 'enemy' alongside Anglo-American occupiers – the aforementioned Iranian/Persian, Salafi/Wahhabi, and Zionist threats to an Iraqi sense of sovereignty. Web 2.0 tools and the discord across various factions and movements saw a further proliferation of perspectives. Many increasingly were in response to the robust videos produced by Al-Qaida and fellow ideological travellers who had focused on recruitment and training while developing sophisticated media arms over the previous decade. Iraqi nationalists, Sufi Islamists, Shi'i militias and political parties, and many of Iraq's minority groups – now armed and organised for communal defence following

American 'liberation' – increasingly turned to media as a vehicle to express political and social grievances as well as to support political group affiliations and identities.

The means of delivery – predominantly mobile phones – also impacted the formulation of content. Short clips, often less than a minute in duration, required intense footage of a unique and often highly localised context, in order to achieve the desired attention; indeed, the filming of Saddam Hussein's hanging, which was 'viral' online globally and on mobile phones across the region, took up just over one megabyte. While American videos veered to the seemingly mundane intimacy of family-focused content or the viral war pornography discussed above – Iraqi and transnational *jihadi* efforts embedded their political content in materials they believed would reach a wider audience. Whether through news programming on the Arab satellite channels – who were willing to include such footage in their coverage – or that would elicit viral status amongst a youthful Arab population now increasingly accessible throughout the mobile networks crossing the region,[45] the phenomenon of transmitting across 'trusted' socially connected networks provided an authority to messages that often undermined official narratives. Cultural norms were altered rapidly as Arab political commentators and even politicians were forced to address – often on live television – issues raised by such clips or by callers and SMS text messages referring to them.

By mainstreaming 'viral' video clips, pictures, ideas, and debates moved rapidly beyond Iraq itself – thereby having the conflicts within Iraq impact the wider region in a fashion and with a granularity not matched by American media whether in its coverage of the conflict or of its own service members' creative output.[46] The delivery along 'trusted' social networks not only elicited a more uncritical eye from the recipient, but also a more immediate and emotive reaction. Regional memes – the supposed increasing number of converts to Shi'ism in majority Sunni states, the rise in Shi'i political action (often perceived as a threat), the increased public hostility towards gays following the viral dispersal of a group of Iraqi men dancing with each other being condemned by senior clerics, and the overwhelming opposition to longstanding police brutality following the mobile phone capture of a young Egyptian man being abused while in custody have all had immense impacts across the entire region. All such events are highly de-contextualised from their original taping, are easy to distribute across mobile SMS networks due to their small file size, require little attention beyond the sensational footage, and almost singularly are emotive in their impact. The impacts of this 'democratisation of information' across highly authoritarian and previously information-starved societies remain unclear. While they clearly engage recipients and draw people into conversation and perhaps political action – in part as a result of the sort of 'politics of aesthetics' that references the politics of the subject noted by Ranciere – they also allow for erroneous speculative ideas and analyses to proliferate due to their de-contextualised orientation. This disassociation from surrounding events clashes with the trusted means of delivery for both Iraqi and American media creators. That both pools of media output are open to a global audience, especially as the lack of authoritative contextualisation is not present outside such networks, proposes that the clips may be seen as a 'differing' as opposed to an 'othering' of political and social differentiation, as the global mediascape's repertoire, while not homogenous, increases in sometimes subtle similarities.

NOTES

1. Iraqis had had precarious and predominantly state-controlled access to the digital age prior to 2002 with the limited exceptions in the northern environs under the control of the Kurdistan Regional Government (KRG). Iraq was estimated to have less than 25,000 internet-enabled personal computers and access to the internet, as well as satellite television and mobile telephony, was highly restricted through state controls as well as the lack of pertinent infrastructures.

2. With the emergence of Al-Jazeera and other privately owned media a potentially revolutionary change has occurred across much of the region with the proliferation of satellite television, access to the internet and mobile phone technologies all obviating the heretofore state-controlled monopoly on information. On Al-Jazeera, the rise of Arab satellite media, and Arab debates over the Iraq war see: Marc Lynch, *Voices of the New Arab Public: Iraq, Al-Jazeera, and Middle East Politics Today* (New York: Columbia University Press 2006).

3. Iraqi foreign intelligence murdered Iraqi and foreign reporters on numerous occasions and Iraqis found cooperating with foreign media by the regime were persecuted by the regime's security establishment. Hazhar Aziz Surme, 'Baathist psychological operations', *The Kurdish Globe*, 9 April 2008; Eason Jordan, 'The News We (CNN) Kept To Ourselves', *The New York Times*, 11 April 2003, available at <http://www.nytimes.com/2003/04/11/opinion/11JORD.html>.

4. Lawrence Wright, 'Web of Terror', *The New Yorker*, 2 Aug. 2004, available at <http://www.newyorker.com/archive/2004/08/02/040802fa_fact>.

5. Mark Glaser, 'Your Guide to Soldier Videos from Iraq', *MediaShift*, 1 Aug. 2006, available at <http://www.pbs.org/mediashift/2006/08/your-guide-to-soldier-videos-from-iraq213.html>.

6. James Der Derian, *Virtuous War: Mapping the Military-Industrial-Media-Entertainment Network* (New York: Routledge 2009) p. 249.

7. Ibid., p. 250.

8. Arthur Kroker, *The Will to Technology & The Culture of Nihilism: Heidegger, Nietzsche, & Marx* (Toronto: University of Toronto Press 2004) pp. 39–40.

9. Ibid.

10. Der Derian, *Virtuous War* (note 6) p. 251.

11. In addition to *Virtuous War*, see also James Der Derian, *Critical Practices in International Theory: Selected Essays* (New York: Routledge 2009); James Der Derian, *After 9/11* (Udris Productions 2003); *Human Terrain: War Becomes Academic* (2010), produced and directed by James Der Derian, David Udris and Michael Udris, Bullfrog Films; James Der Derian, *Antidiplomacy: Spies, Terror, Speed, and War* (Oxford: Blackwell 1993).

12. Der Derian, *Virtuous War* (note 6) p. 251.

13. See Arjun Appadurai, *Modernity at Large: Cultural Dimensions of Globalization* (Minneapolis: University of Minnesota Press 1996).

14. Michael Y. Dartnell, *Insurgency Online: Web Activism and Global Conflict* (Toronto: University of Toronto Press 2006) p. 5.

15. Rachel Hughes, 'Through the Looking Blast: Geopolitics and Visual Culture', *Geography Compass* 1/5 (2007) pp. 976–994.

16. The so-called 'highway of death' is Highway 80 which runs between Kuwait City, Kuwait and Basra, Iraq. Iraqi forces fleeing Kuwait were heavily bombarded in 1991 by US and UK forces. This

proved controversial as the attacks were cited as strong examples of the unnecessarily disproportionate use of force, on troops and refugees, who were at the time retreating from the theatre of operations.

17. The notion of 'performances of nationhood' is used as developed in Michael J. Shapiro, *Methods & Nations: Cultural Governance and the Indigenous Subject* (New York: Routledge 2004).

18. Der Derian, *Virtuous War* (note 6) p. 251.

19. On the varied audiences of the video-releases of Al-Qaeda see, Mark Lynch, 'Al-Qaeda's Media Strategy', *The National Interest* (Spring 2006) pp. 50–56, available at <http://www.nationalinterest.org/Article.aspx?id=11524>.

20. See the MTV produced documentary "Iraq Uploaded" and the award-winning documentary "The War Tapes" depicting the deployment of the New Hampshire National Guard to Iraq – a film shoot entirely with footage shot by the unit itself while deployed to Iraq.

21. See Mark Glaser, 'Porn Site Offers Soldiers Free Access in Exchange for Photos of Dead Iraqis', *USC Annenberg Online Journalism Review*, 20 Sep. 2005, available at <http://www.ojr.org/ojr/stories/050920glaser/>; as well as the article 'Foto dell'orrore in Iraq in cambio di porno', *Il Corriere della Sera*, 26 Sep. 2005 (a mainstream Italian daily).

22. One of the 'five pillars' of Islam the *hajj* refers to the pilgrimage every able Muslim must attempt to make to Mecca within their lifetime. Muslims who complete the pilgrimage are referred to as *hajji* to denote their observance of religious fidelity. Many US forces in Iraq have adopted the term in reference to all Iraqis as a seemingly neutral term acceptable to image-conscious officers in spite of its pejorative connotations to many who use it in such a fashion.

23. *Combat Diary: The Marines of Lima Company*, directed by Michael Epstein, produced by Viewfinder Productions (2006), available at <http://www.aetv.com/listings/episode_details.do?episodeid=163325>.

24. Brian Lowry, 'Combat Diary: The Marines of Lima Company', *Variety,* 23 May 2006, available at <http://stage.variety.com/review/VE1117930623?categoryid=32&cs=1>.

25. Ibid.

26. A portion of *Combat Diary: The Marines of Lima Company*, focuses on the introduction of some Iraqi special forces who then work alongside Lima Company in operations.

27. See: Douglas Little, *American Orientalism: The United States and the Middle East since 1945*, 3rd ed. (Chapel Hill, NC: The University of North Carolina Press 2008); Melani McAlister, *Epic Encounters: Culture, Media, and U.S. Interests in the Middle East Since 1945*, 2nd ed. (Berkeley, CA: University of California Press 2005); Karim H. Karim, *Islamic Peril: Media and Global Violence* (London: Black Rose Books 2000); Thierry Hentsch, *Imagining the Middle East*, trans. by Fred A. Reed (New York: Black Rose Books 1992); Edward Said, *Covering Islam: How the Media and the Experts Determine How We See the Rest of the World* (New York: Vintage Books 1981); Edward Said, *Orientalism* (New York: Vintage Books 1979).

28. Michael J. Shapiro, *Violent Cartographies: Mapping Cultures of War* (Minneapolis: University of Minnesota Press 1997) pp. 47–49, passim.

29. Ibid., p. 47.

30. Ibid., p. 48.

31. Three aspects emerge that comprise the 'Web 2.0' generation of software applications and clearly alter social relations, networks and behavioural interactions both between users and between users and the technologies themselves: (1) Rich Internet Applications (RIA) such as flash or Ajax (software making the app work in a browser) are often iterative as many users can constantly update the application (2) Service Orientated Architectures (SOA) such as feeds like RSS, Web Services allowing sharing of user-generated content such as Flikr and YouTube, and finally the ability to compile "Mash-Ups" – bringing such content together in unique and shareable formats all emerge in a (3) Social Web that encourages interaction with the end user (creator, audience and participant) by allowing the tagging of content, and its dissemination and organisation across wikis, blogs, and podcasts.

32. See: Misha Glenny, *McMafia: A Journey Through the Global Criminal Underworld* (New York: Knopf 2008); John Robb, *Brave New War: The Next Stage of Terrorism and the End of Globalization* (Hoboken, NJ: John Wiley & Sons 2007); Moises Naim, *Illicit: How Smugglers, Traffickers, and Copycats are Hijacking the Global Economy* (New York: Doubleday 2005).

33. Der Derian, *Virtuous War* (note 6) p. 251.

34. See Benjamin J. Muller and John H. W. Measor, 'Securitizing the Global Norm of Identity: Biometrics and *Homo Sacer* in Fallujah', in Benjamin J. Muller, *Risk, Security, and the Biometric State: Governing Borders and Bodies* (London: Routledge 2010).

35. Destroyed in part in the winter of 2007.

36. The Iraqi Ba'th Party ruled Iraq from 1968 until their overthrow by US military forces in March–April 2003. The competition for what were previously Ba'th-co-opted symbols an imagery by neo-Ba'th, Iraqist nationalist and Islamist political groups and movements make the performance of the Iraqi nation a very much fluid proposition within the insurgency. On Ba'th imagery see: Eric Davis, *Memories of State: Politics, History and Collective Identity in Modern Iraq* (Berkeley: University of California Press 2005); and Kanan Makiya, *The Monument: Art and Vulgarity in Saddam Hussein's Iraq* (London and New York: I. B. Tauris 2004).

37. Referring to the al-Aqsa mosque in Jerusalem.

38. The K'aba is the holiest site in Islam located inside the Masjid al-Haram mosque in Mecca. Islamist imagery has long judged the 1990s US military presence in, and long-standing alliance with, the Kingdom of Saudi Arabia as an affront and tacit occupation of the holy cities of Islam by infidel forces. Islamist groups including al-Qaeda commonly advanced such arguments throughout the 1990s. See Michael Bonner, *Jihad in Islamic History: Doctrines and Practice* (Princeton, NJ: Princeton University Press 2006).

39. See *Millennium: Journal of International Studies* 34/3 (2006).

40. Jacques Ranciere. The Politics of Aesthetics, trans. by Gabriel Rockhill (New York: Continuum 2006) p. 51.

41. See Jacques Ranciere, *Disagreement: Politics and Philosophy* (Minneapolis: University of Minnesota Press 1998); Jacques Ranciere, *The Nights of Labour: The Workers Dream in Nineteenth Century France* (Philadelphia: Temple University Press 1989); and Ben Davis, 'Ranciere for Dummies', *Artnet Magazine* (2007), available at <http://www.artnet.com/magazineus/books/davis/davis8-17-06.asp>, accessed 26 Jan. 2007; contrast with Louis Althusser, 'Lenin and Philosophy and Other Essays', *Monthly Review Press* (1972).

42. Appadurai, *Modernity at Large*, p. 35.

43. One aspect evident across the political spectrum found within, and cutting across social class divisions within the US military, is an increasing sense that those serving in the military are more honourable than many of those they serve. First addressed by Andrew Bacevich, such a phenomenon was consistent within the video products emerging out of Iraq. As much or more than focusing on the Iraqis surrounding them, American forces evidenced concern with their fellow citizens, their insularity from the combat experience, deprivations of service, and general disengagement from the 'mission' they themselves were sacrificing so much to pursue. See: Andrew J. Bacevich, *The New American Militarism: How Americans Are Seduced by War* (New York: Oxford University Press 2006).

44. On the surge see: Thomas E. Ricks. *The Gamble: General David Petraeus and the American Military Adventure in Iraq, 2006–2008* (New York: The Penguin Press 2009); and Peter R. Mansoor, *Baghdad at Sunrise: A Brigade Commander's War in Iraq* (New Haven: Yale University Press 2008). Beyond the immense efforts of US, Iraqi, and allied forces in dampening levels of violence in Baghdad, Ricks makes plain the strategic failure at the political level as Iraqi factions moved not one iota forward on the issues dividing them. The success – encapsulated in the 'victory' narrative at Washington salons and on cable news outlets – belied a sense of stasis described by many US forces members on the ground through their communications home as viewed by the authors.

45. Arab youth demographics represent an enormous bulge in population and political actors – and economic actors – are increasingly attempting to reach such a dynamic group. Finding employment, social cohesion, ecological impact, and political mobilisation are only some of the important factors researchers are concentrating on in examining this rapidly advancing cohort. For the Arab Human Development report(s) detailing such challenges in 2005 and 2009 see <http://www.arab-hdr.org/>; for the Human Rights Watch report 'False Freedom – Online Censorship in the Middle East and North Africa', see <http://www.hrw.org/en/reports/2005/11/14/false-freedom>.

46. This emphasises how viral videos and new media contribute in altering distance and place through speed and Web 2.0 technology, and, consequently, how war is waged and mediated on and beyond the battlefield. See: F. Macdonald, K. Dodds, and R. Hughes, *Observant States: Geopolitics and Visual Culture* (London: I.B. Tauris 2010).

"Those About to Die Salute You": Sacrifice, the War in Iraq and the Crisis of the American Imperial Society

FLORIAN OLSEN

School of Political Studies, University of Ottawa, Ontario, Canada

This exploratory article argues that the mobilisation process for the war in Iraq has revealed and exacerbated the fault lines across American society and brought on a crisis in army-society relationships in the US to which the state has ultimately responded by announcing it would withdraw troops from Iraq. To explore these tensions, I present the analytical framework of "American imperial society" as an alternative problematisation of 'empire' and hope to show its utility to highlight how the repercussions of the US's political ambitions abroad are felt far beyond the battlefield. In doing so, it also provides the first account of empire which builds on the sociology of Pierre Bourdieu. I finally explore these questions by looking at the debate taking place over the meaning and legitimacy of American military sacrifice in three major American newspapers and two magazines.

The mass mobilization of the reserves and the National Guard has created neither widespread public support nor opposition to the war. This is because the troops who have borne the burden of the Iraq War have been disproportionately drawn from small-town America and the inner cities, *not* from those social groups who shape national policy.[1]

— Charles Moskos

INTRODUCTION

If we are to believe Charles Moskos's quote in the epigraph, as of 2005, the American population was meeting the war in Iraq with relative indifference.

This article argues quite the contrary. The highly charged debates in newspapers, think-tank contributions to the public sphere, congressional commissions and the resurgent and extremely polemical academic discussion on the costs and consequences of "American empire"[2] are only some of the symptoms which reveal how the mobilisation process for the War in Iraq has exacerbated the fault lines across American society. I further suggest it has brought on a crisis of the American imperial form of society: a crisis for both the US geopolitical domination and hegemony[3] in the society of states and for the American society, leading to an intense conflict between citizens over the war to which their state[4] has ultimately responded by announcing it would withdraw troops from Iraq. To explore these tensions, this exploratory article presents the analytical framework of "American imperial society"[5] as an alternative problematisation to 'empire' and hopes to show its utility to highlight how the repercussions of US political ambitions abroad are felt far beyond the battlefield. In doing so, it also provides the first account of empire which builds on the sociology of Pierre Bourdieu.[6]

Imperial societies dominate the system of states as well as subordinate populations within their borders. To capture this dual dimension of imperial rule, the framework introduces a further concept, "the field of citizenship," a continuum along which populations can be graded according to the different degrees of inclusion and exclusion to which they are submitted. The imperial society and field of citizenship framework build on the contributions of postcolonial scholarship while also attempting to overcome some of its limitations by highlighting differing gradations of imperial rule, including class relations.

I explore relations of imperial rule through the question of military service which spans the divide between domestic and foreign politics. Soldiers are the ultimate instruments through which a state imposes its geopolitical domination on the international society. Simultaneously, military service has been the traditional hallmark of first-class citizenship and thus of dominance within the *field of citizenship*. In this sense, the 'imperial society' framework displaces our interrogations from traditional relations of geopolitical domination to the more insidious forms of geopolitical dominations that take place *within* state borders, the political body or *field of citizenship*: those places where belonging and exclusion or foreignness are ultimately negotiated and decided.

The article begins by outlining the concept of "imperial society" and brief notes on the cultural sociology of Pierre Bourdieu. The second section of this paper provides a short overview of the relationship between military service and citizenship in the US to contextualise contemporary tensions arising out of the War in Iraq. The third section then looks at how the war in Iraq has ignited tensions over the distribution of military sacrifice and polarised the political field. I explore these questions by looking at the debate taking place over the meaning and legitimacy of American *military sacrifice* in three major American newspapers and two magazines. The final section of this paper looks at how the American state has responded to the acute polarisation of the political field and the exhaustion of the armed forces.

FIRST- AND SECOND-CLASS CITIZENS IN THE AMERICAN IMPERIAL SOCIETY

An Imperial Society

This study builds on historian Christophe Charle's[7] model of the "imperial society" adapted from his comparative social and cultural history of the French, German, and British empires of the early twentieth century.[8] Unlike the other empires of their time (i.e., Russia, Austria-Hungary, the Ottoman Empire), imperial societies also enjoy the means to exert a near unparalleled linguistic, cultural, economic, political and scientific dominance on international society. They fiercely engage with other societies in a contest for global preeminence in the name of "manifest destiny." Contrary to "ordinary nations" the imperial society claims "to be a universal cultural model with the vocation to draw into its orbit less autonomous nations and peoples."[9] This is what I understand hereafter by the imperial society's logic of *external* geopolitical domination.

While imperial societies enjoy the means to aggressively exercise their influence in the system of states, this article is most interested in the 'domestic' repercussions of this quest for global pre-eminence in relations between dominant and dominated groups in American society. Traditionally empire is defined as the annexation of *foreign* lands or peoples. This definition of imperialism as geopolitical domination rests on the modern political practices and understandings which tie citizenship primarily to territory, and in particular to the nation-state from which marks the modern resolution of the problematic relationship between community and territory.[10] Postcolonial scholars have demonstrated that foreignness and citizenship are largely the results of cultural and political practices which designate certain populations as belonging to the community and excluding others who lack the proper characteristics: race, national origin, gender, political ideology or religion to warrant inclusion.[11] Though one of the primary sites which construct citizenship and belonging, geopolitical borders also formalise and reinforce, in this process, deeper rooted cultural understandings of *community*. "Foreign" crystallises a double meaning which draws our attention to the cultural and political borders of citizenship as much as it does the juridical and territorial borders of states.[12] David Campbell and Amy Kaplan both remark that "liminal" populations[13] in the US – African Americans, Amerindians, Hispanics, and even politically and sexually marginal citizens – were made into 'foreigners' through an effect of *geopolitical refraction*; they were identified in relation to a "homeland *elsewhere*"[14] to which their 'heart' was held to belong (irrespective of whether or not this is true). For instance, in the wake of the 1898 Spanish-American War, the US Supreme Court had determined Puerto Ricans to be "foreign in a domestic sense",[15] exemplifying the US's long tradition of cultural and racial, not simply territorial, citizenship. Indeed, even after having received *equal status* under the *law*, dominated minorities remain second-class citizens, relegated to dominated positions and subject to a form of domestic colonialism: racism, scapegoating, and racial discrimination.[16] Thus the United States qualifies as an 'empire' *before we even begin* to look at its 'foreign relations' or its global politics.

But if postcolonial theorising helps us expose imperial rule *within* the United States by questioning the foreign/domestic divide, however, it risks amalgamating the practices of imperial exclusion into a one size fits all model. Kaplan[17] rightfully notes that Diasporas have ties to both their homelands of origin and to their country of adoption, a complexity the state then attempts to discipline.

> [The notion of Homeland] does this not simply by stopping foreigners at the borders, but by continually redrawing those boundaries everywhere throughout the nation, between Americans who can somehow claim the United States as their native land, their birthright, and immigrants and those who look to homelands elsewhere, who can be rendered *inexorably foreign*. This distinction takes on a decidedly racialized cast through the identification of the homeland with a sense of racial purity and ethnic homogeneity, which even naturalization and citizenship cannot erase.[18]

Kaplan's argument is a staple of the poststructuralist critique of the metaphysics of presence I call the self/other binary. Canonical French Theorists like Edward Said or Judith Butler[19] argue that difference not only precedes identity, but stabilises it. Indeed, we would mislead ourselves to say: "I think therefore I am"; in reality, "I am myself because I am *not* you." Whether it is through "Orientalism," cartographic practices,[20] or "foreign policy,"[21] identity attempts to stabilise itself by rejecting other forms of subjectivity. The West "created the Orient", in Said's famous formula, but it was really cementing its own identity. In another example, Anne-Laura Stoler writes: "That colonial states policed and protected the privileges bestowed on some by making them police the moral values, familial forms, and political affiliations of those within their communities and of those suspected of really *being others* are echoed in social vigilantism today."[22] Here again, Stoler highlights how "otherness" is a central component of both the constitution of identity and imperial practices of inclusion and exclusion.

The point here is that while postcolonial theorists are correct to point out that identity (self) does not exist on its own, apart from difference, and that states and empires thrive on producing "others," *excluded citizens*, there is a risk that the *didactic point* overshadows the complexity of the real practices of inclusion and exclusion it is meant to highlight. While it is a starting point for understanding inclusion and exclusion, the self/other binary cannot, without further analysis and empirical reconstruction, adequately grasp the complexity of these practices. David Campbell perceives the risk in this and writes: "The demarcation of self and other is, however, not *a simple process that establishes a dividing line between the inside and the outside*. It is a process that involves the gray area of liminal groups in a society, those who can be simultaneously self and other – outsiders who exist on the inside."[23] However he soon goes on to argue: "The strategy for dealing with liminal groups was to *identify them with the foreign, usually the enemy.*"[24] So, like in the above ruling by the Supreme Court ("foreign in a domestic sense"), for all his best intentions, Campbell leaves us with a dichotomous account of citizenship and identity. The danger is that this very effective analytical formula has become clichéd and may stand in place of an analysis of the *actual practices* of inclusion and exclusion employed by any one group, state or empire. To take just the example of the United States, an analysis like Campbell's misses the fact that liberal societies are not primarily involved in producing and stigmatising difference. An important part of how cosmopolitan-liberalism operates is precisely by assimilating difference, not rejecting it. They produce *sameness*, and therein lies the claim to legitimacy of cosmopolitan empires: they include, not the opposite. Furthermore, a cursory study of citizenship laws or cultural attitudes toward minorities in the United States will reveal that "liminal groups" were not all thrust into one category of "otherness" or "foreignness." Thomas Jefferson certainly held American Indians in higher esteem than Africans, and the Supreme Court treated these groups differently. As historian Linda Kerber[25] writes, US citizenship has been made up of multiple "braids" corresponding to the different populations residing in what would become American territory, each beholden to occupy a different place in the racial hierarchy and obtaining rights at very different times. Lastly, the self/other dichotomy fails to capture the mobility of dominant and dominated groups over time, as their reserves in capital holdings increase or decrease, as a result of political changes or successful struggles for recognition.

One of the ways to overcome the problem posed by the self/other binary is to look at imperial rule along a continuum of inclusion and exclusion within a *field of citizenship*. For Pierre Bourdieu, the "global volume of capital" or wealth individuals, social classes,[26] class fractions and groups possess, as well as their relative shares of more specific cultural, social, political and symbolic "assets,"[27] determine their positions of subordination or dominance on a map he terms "social space." On this map, the *field* represents narrower relations of dominance within more specific terrains of struggle upon which social actors, groups, classes and class fractions act and compete to augment or reproduce shares of influence or capital specific to this field. The *field of citizenship* spans from the complete exclusion of 'foreigners' ('otherness' in the most traditional sense) to the inclusion of full or first-class citizens, with median, marginal or *borderline* positions occupied by women, coloured women, racial and sexual minorities, and dominated class fractions.[28] At the heart of the contests within this field lies the competition for individuals and groups to gain recognition or inclusion in the body politic. This concept sports the immediate advantage of systematising the obvious points that racial and ethnic minorities inhabiting the US territory or even enjoying nominal(/legal) citizenship may be considered foreigners and prey to a form of domestic or 'internal' colonialism. But it will also help us ask the question: does second-class citizenship for some lower-class white males not correspond to a dynamic of imperial domination by dominant class fractions? In this sense, the 'imperial society' model provides a more nuanced theoretical framework to explore the impact of traditional relations of geopolitical domination on the (geo)politics that take place *within* state borders, the political body or *field of citizenship*. This is an especially pressing move for the study of war and military conflicts which engage both a society's reach into the system of states or international society and exact a tremendous cost on the political and social structures within its borders.

Citizenship, Class and Military Service in the United States

Military service for white, young adult males has been historically one of the principal cultural practices of conferring and policing citizenship: at once defending the geographical borders of the state from invaders *and* the (geo)cultural and (geo)*political* traits of the nation or *polis* through the exclusion of second-class citizens. Military sacrifice constitutes one of the fundamental political obligations,[29] but also one of the privileges,[30] of membership in the republican community. The citizen-soldier devotes his life to the state because its fundamental law, or constitution, and the principles on which they are founded help him to lead a life of virtue.[31] But women, racial minorities, paupers, immigrants and others were excluded from equal standing in the community[32] because they were deemed "by nature or condition or conviction, incapable of the virtues good citizenship requires."[33] Similar justifications were given to exclude African Americans from fully participating in the army, especially in peacetime, or to organise them into segregated units.[34] The citizen-soldier tradition has profound cultural ramifications for the establishment of gender roles[35] in the United States, notably by excluding women from combat. In 1968, the US District Court, Southern District of New York in *U.S. v. St. Clair* upheld the constitutionality of the sexual division of labour, stating that *"Congress followed the teachings of history that if a nation is to survive, men must provide the first line of defense while women keep the home fires burning."*[36] In the end, the relationship between military service and republicanism was based on the universal requirement of service for members of the polity, excluding those whom it casts to the margins of the field of citizenship.

The end of the Second World War brought sea-changes to the American imperial society's field of citizenship. Segregation in the armed forces collapsed as America faced war in Korea, and women gained a limited right to integrate into the armed forces.[37] The Gates Commission abolished conscription in 1973, and the all-volunteer army replaced the 'citizen' army of old. In the postwar context of neo-liberal "re-regulation",[38] military service became a career choice. Though today, the neo-liberal logic of sacrifice frames military service as an individual choice conditioned by neutral market processes rather than the result of systemic inequalities; in reality, the institution of the all-volunteer force reinforced many of the inequalities of racial and socioeconomic status it was supposed to correct.[39]

Today, less than 1% of the American population serves in the military.[40] "As in Canada and other advanced capitalist nations with voluntary military service, recruits in the US are spatially as well as socially marginal citizens," notes Cowen.[41] They are disfavoured both in terms of the economic capital (wealth) and cultural capital (education) they possess. Though most service members must hold a high school diploma in order to enlist, thus making them 'more' educated than the comparative civilian population,[42] Bachman, Segal, Doan et al. and Tripp[43] demonstrate that access to cultural capital in the form of college education or desire to enrol in college separates those likely to volunteer from those who won't. The requirement to hold a high school diploma for a career in the military effectively excludes from service those class fractions most dominated in social space and the field of culture. Indeed, larger concentrations of cultural capital, usually university diplomas, grant access to white collar jobs and thus the middle classes.[44] For this reason, recruitment most effectively targets the dominant fraction of the dominated classes (the highest strata of the lower classes and the lowest strata of the middle class): racial minorities and white citizens from economically disadvantaged, rural areas or inner cities (see Beeghley for a portrait of the class structure in America).[45]

These enlistment patterns reflect salient inequalities in the labour market and in the imperial society's field of citizenship. As Bourdieu[46] shows, the global volume of capital an individual or social group possesses, and the number of generations (thus length of time or 'age of capital') it has been in their families, statistically measure the *"modal trajectory,"* or "class trajectory", upon which members of the social group are most likely to follow from "the field of possibilities" open to them. Bourdieu terms "habitus" these sets of practical schemes and the strategies individuals and groups employ to meet the unique challenges that arise from their position in social space, whether it be one of dominance or subordination.[47] In the American imperial society, the horizon of possibilities that the unequal reserves of capital holdings and thus the very different habituses of dominant and dominated groups have largely determined whether they will consider service in the armed forces.[48] Most affluent citizens can avoid military service. For one, neither political obligation nor legal requirement constrains them to fight.[49] And their respective pools of cultural, economic and social capital open a wider horizon of life choices and professional strategies than dominated groups who cannot count on the same assets. Because of the "homologies between the fields,"[50] one's place in the field of classes simply reproduces in economic and educational terms (economic and cultural capital) the boundaries that separate different castes of Americans within the field of citizenship, and inversely. Each class's modal trajectory is structured by their respective places in the field of citizenship and the specific obstacles the imperial society places in their way.

With one major exception, that of African Americans (see below), the structure of casualty distribution in Iraq reflects the above inequalities. The wars in Afghanistan and Iraq have exacted a terribly high price on small-town and rural areas, especially in the "rural South and rural West".[51] The former geographical region provided the armed forces with four out of every ten new service member in the year 2000.[52] One year after the invasion of Iraq, nearly one in two soldiers killed in the conflict hailed from small-town America, that is "communities with populations under 20,000."[53] This brought the death toll of these communities "60 percent higher than the death rate for those soldiers from cities and suburbs".[54] The much larger price paid by these communities reflects class divisions stratified in geographic space, notably between rural and urban areas, between neighbourhoods which promise very different prospects for employment, and increased exposure to violence and poor living conditions.[55] Opposite this, cities and especially metropolitan areas also facilitate the acquisition and consumption of elite cultural capital because of the presence of museums, art houses, operas, libraries and major universities.[56] While the South is known by military sociologists for its "military tradition",[57] rural areas have also been deeply affected by deindustrialisation and outsourcing since the 1970s further adding to those regions' economic precariousness.[58] To this degree, geographically and economically marginal communities and citizens also occupy borderline or dominated positions in the field of citizenship through the homologies in the American imperial society.

Further, racial inequalities remain salient in the all-volunteer army. Though they make up just under 13 percent of the civilian work force over seventeen years old, African Americans represent 20 percent of active duty forces; however, contrary to Hispanics, beginning in the 1980s, they were purposefully drawn away from combat intensive military occupational specialties,[59] accounting for much lower African American casualties in the Post-Vietnam army, and in Iraq[60] specifically, than their weight in active forces would suggest. Though blacks are no longer disproportionately being sacrificed on the battlefield, this role is now devolving to Hispanics, as Gifford suggests, who were disproportionately killed during 'conventional combat phases' of the war.[61]

Profound transformations in the imperial society's field of citizenship, politics and ideological production brought on by desegregation, the civil rights movement, and the rise of the volunteer army have all deeply affected the collective understanding of military service as a choice, the demographic composition of the armed forces, and the changing nature of the social divides mirrored in army-society relations. Furthermore, the casualty structure of the Iraq War reflects inequalities between urban and rural areas and class divisions within the white population which are especially visible as differences in cultural capital or college propensity. Unequal holdings in economic, cultural and social capital contribute to shape the real life possibilities and strategies (habitus) to which individuals can aspire. Because of the homologies between the fields and the struggles within them, groups dominated in one area of life tend to be marginalised in the next. This means that unequal opportunities stratified by race or class can be theorised as more fundamental inequalities in the imperial society's field of citizenship. Military service has increasingly become the hallmark and function of the imperial society's second-class citizens. These citizen-soldiers bear a disproportionate obligation to sacrifice to secure the US empire's geopolitical domination abroad in order to receive basic privileges other well-to-do citizens take for granted.

THE CRISIS OF THE AMERICAN IMPERIAL SOCIETY

Debating Sacrifice in Iraq

Already faced with the military occupation of Afghanistan, the War in Iraq confronts the American public with an enormous toll of military casualties now surpassing 4,000 killed and 30,000 wounded.[62] If the War in Iraq has forced the American imperial society to renegotiate its place of symbolic and political dominance in global politics, it has equally strained relations between army and society, first- and second-class citizens, a fact reflected by how the burden of war has weighed very differently on both dominant and dominated class fractions. I call "crisis[63] of the *imperial* model of society" this simultaneous renegotiation of US dominance in global politics and debate over the relationship between military service and citizenship domestically. Geopolitical conflict between the US and subordinate societies is mirrored by political conflict between first- and second-class Americans within the imperial society's field of citizenship.

But by 'crisis' of the American imperial society I mean nothing like imminent revolutionary transformation, but rather, as Stuart Hall argues, a social and political process through which social tensions in the form of class competition escalate to a point where the state must decisively intervene to mediate them. Appears first a 'narrative' of crisis[64]. Social and political actors sow webs of meaning between disparate events creating narratives of causality which purportedly illuminate or 'explain' deeper social unrest or anxieties. Appeals to crisis question the legitimacy of the social order and pretend to offer a solution to the problem. To augment their shares of political capital and obtain dominance in the political field, social actors accuse their political adversaries (the 'guilty'), deplore the misery of their victims, and identify a new guard of social actors capable of redressing the situation.[65] Citizens are thus called upon to identify to one or the other of these three roles in relation to their experiences of the social world. Brought on by the growing contestation of the social inequalities revealed by the war, the crisis reverberates and manifests first, keeping with Hall, as cultural anxieties within the political, cultural and ideological fields in the form of "moral panic." It is thus a crisis in "hegemony" or of the cultural mechanisms which secure political consent. But it is only through the process by which the state directly intervenes to bring an outcome to class conflict that we can say that social tensions have escalated to or near the point of 'crisis.'

I argue the war in Iraq has exposed the fissures at the heart of the field of citizenship in a way it has not done in decades. Newspaper debates[66] over military sacrifice in Iraq furnish excellent quantitative and qualitative evidence that testify to this polarisation of the political field. I examined over two hundred articles from *The New York Times*, *The Wall Street Journal*, *USA Today*, *Newsweek* and *The Nation*, and focused on the period extending between 11 September 2001 and 4 November 2008 – date of President Obama's election. A similar newspaper string search conducted for the period between 1 January 1992 and 1 September 2001 returned only twenty-nine articles in the *New York Times*, *USA Today*, *Newsweek* and *The Nation*, barely one third of the documents.[67] A quick glance at the content of these articles showed that almost none of them had relevance to debate over army-society relations in the US. "Sacrifice" was metaphorical, related to feature films or ongoing conflicts in the world. The sample for the period between 2001 and 2008 showed significant polarisation of the political field that directly touched the divisive relationship between sacrifice and citizenship.

To explain this let us consider first the US public has now been at war for nearly eight years on two fronts. The Wars in Iraq and Afghanistan mark the imperial society's longest military engagements since the Vietnam War, thus over the three and a half decades since the fall of Saigon. In the aftermath of 9/11, support for reinstating conscription ran high in the American population. Seventy-six percent of those interrogated consented to this option if "it [became] clear that more soldiers are needed in the war against terrorism." By the summer of 2005, this support had plummeted to 27 percent.[68] The prospect of continued sacrifice no longer appealed to the American population on a personal or collective level. Opinion[69] over the War in Iraq has steadily eroded, especially amongst African Americans. Between 2000 and 2007, the number of black recruits plummeted a striking "58 percent", as well as their overall support of the War, marking a clear change in the African American habitus. *The New York Times* reported that African Americans opposed the war not only in majorities upward of 70 percent, but at almost double the numbers of white Americans.[70] Furthermore, debate over the continuation of the conflict figured prominently in the political field and set the tone in both the 2004 and 2008 presidential elections, as well as in the mid-term 2006 congressional elections during which dissatisfaction over the war in Iraq enabled the Democrats to gain control of Congress.[71] Not only the first black presidential candidate in US history, Barack Obama was also one of the first politicians of major notoriety to have opposed the war from the onset,[72] setting him apart from both his Democratic and Republican contenders, Hillary Clinton and John McCain, and increasing his share of political capital. American politics of the last decade have been defined by Iraq, a war which remains unfinished; in the 1990s, the US was only at war briefly in 1991 and again in 1998, and still only in an extremely limited manner. In this fundamental sense, the political field has been shaped by polarisation over the war in a way it could not be a decade earlier.

While never a serious issue in presidential politics, the question of reintroducing the draft remained, beyond the collapse of public opinion in favour of this option, the subject of great public debate in the political field taken largely: newspapers, think tanks, within government and amongst politicians such as New York Congressman Charles Rangell. To a degree, it mirrored a larger divide in American society over the utility of the war in Iraq and fault lines over the meaning of citizenship and the respective obligations of citizens. As early as 2003–2004, while soldiers began to resign themselves to serving multiple combat tours in Iraq, journalists, editorialists and readers argue that the conflict in Iraq has given birth to two Americas.[73] Though casualties were still relatively low, a year and a half into the conflict, and one month away from the 2004 presidential election, Congressman Charles Rangell, Democrat of New York, wrote that the extensive utilisation of the Reserve Forces and National Guards resembled a "backdoor draft."[74] As the conflict lengthened nearly into its fourth year with both service members and civilian casualties multiplying, a 2006 Christmas editorial in *USA Today* denounced the moral disengagement of the American public from the war effort. "In Iraq, the editorial wrote, the troops are surrounded by hardship and a frenzy of violence. Back home, the frenzy is one of excess: shoppers battling for parking spaces in malls, snapping up everything from flat-panel TVs to the latest video games."[75] On Memorial Day 2007, a *USA Today* editorialist added his voice to other journalists asking for civilians to share a greater part of the burden of war: "The lives of soldiers fighting in Iraq – or headed there, or just returned – have become tapestries of sacrifice not easily fathomed by Americans preoccupied this weekend with barbecues and holiday sales.[76] In October of 2007, as President Bush's nearly 30,000-strong troop surge was underway, an American soldier returning from Iraq denounced "the disparity between the lives of the few who are fighting and being killed, and the many who have been asked for nothing more than to continue shopping."[77] "The city parties on" he wrote in the *New York Times*, "America has changed the channel."[78]

By January of 2008, the disproportionate toll that the war was exerting on small and rural American communities with little opportunity was finding its way into the pages of *USA Today*. In small towns like Lee, Maine, numbering one thousand, the loss of even one, let alone two, young men brought great distress on the locals.[79] News writers and readers argue that American cities are comparatively spared the price of blood because they offer greater prospects for employment. They are filled with colleges and students, bankers, and children of the country's more affluent classes, who feel little real repercussions from the war.[80] Here, news articles and op-eds recognise that greater opportunities to obtain economic capital directly translate into different habituses: class-strategies leading away from military service. These articles thus denounce how the reality of the urban-rural divide of sacrifice is also one of class divisions across geographical space. In this sense, the imperial society's logic of geopolitical domination in world politics exacerbates concern over the geographical subordination of the South and rural areas, and the economically disfavoured citizens which inhabit them to the privileged city-dwellers. They denounce the fact that the unequal obligations in the field of citizenship toward sacrifice are stratified by geography within the very borders of the American imperial society.

The *Nation* and *Newsweek* further accuse "Chickenhawks"[81] like Dick Cheney of asking for sacrifice though they once avoided the draft. While the war squanders America's resources in economic and human capital,[82] the imperial society's elites have given large tax cuts that benefit other members of the dominant class fractions and taxed those already dominated in the economic and cultural fields.[83] That all Americans do not need to sacrifice then raises the question of whether the war is worth fighting at all. *New York Times* editorialist Bob Herbert drives this point home:

> [Winning the war in Iraq] would require implementing a draft. It's easy to make the case for war when the fighting will be done by other people's children . . .. If most Americans are unwilling to send their children to fight in Iraq, it must mean that most Americans do not feel that winning the war is absolutely essential.[84]

But the apparent bipartisan support expressed across many of the newspapers for reintroducing the draft is misleading when taken at face value. It needs to be recast within the logic of debate existing at the time within the political field in the imperial society. Bourdieu[85] writes that each field regulates competition differently, and actors competing within a field must respect a certain set of rules in order to be successful. These rules both reflect the history of the field and the interests of the dominant actors within it. In this case, the political field was dominated by political and social actors close to the George W. Bush administration and by those who espoused their discursive logic. This dominant position equated support of US troops and military families to support for the president's policies in Iraq. Maximum political rewards (political capital) came to those who appealed to the nationalist logic which dominated the political field. So if strategically speaking both opponents and partisans of the war could call for the draft, they did so for dramatically different reasons. In the former case, repeated calls for the institution of conscription enables opponents of the conflict to sap support for the war by threatening the serenity of Americans who live in a "golden bubble" – that is citizens who can continue to go about their daily routines unscathed by the conflict.[86] Interestingly, *The Nation* casts this strategy as one fraught with political danger and categorically rejects calling for the draft in order to generate opposition to the war.[87] But in fact, the *Times* pays only lip service to the republican themes of duty and shared sacrifice in order to trap their political opponents in the contradictions of their own rhetoric: the fact that they support the war but refuse to pay the price that they ask of others. In reality, the antiwar discourse attempts to make its position more acceptable to the general public by appealing to patriotic themes.[88] Again, we must understand this political move in the prevalent discursive context in the political field. The administration and supporters of the war have successfully equated domestic opposition to the conflict in Iraq to condemnation of the soldiers in Iraq and sometimes even support for the 'terrorists':

> I've spent a year in combat in Iraq and have firsthand knowledge of the facts. If you run from Iraq, the radical Islamists will take over that country, and it will become a base of operations for worldwide terror. Then the terrorists will return to Afghanistan in full force, and the casualty count there will increase . . . . All the whining and complaining will not change the facts. We in the military are willing to face the challenges before us to win. *All we ask is that Democrats keep their mouths shut and stop encouraging the enemy to kill us.*[89]

But the republican poll of the debate on sacrifice also counts in its midst supporters of the occupation of Iraq who call for conscription because they envisage it as the only viable strategy for America to win. For example, an "inactive" marine corporal writes in a *Newsweek* column that the draft needed to win in Iraq must be nothing like the

> Vietnam style draft, where men like the current vice president [former vice president Dick Cheney] could get five deferments . . . . No, I am talking about a fair and universal, World War II-style draft, with the

brothers and sons of future and former presidents answering the call (and unfortunately, dying, as a Roosevelt and a Kennedy once did) on the front line.[90]

Partisans of this pro-war republican conception of sacrifice tie their political opponents' lack of support for the war to the loss of traditional values and link them to other upheavals in the fields of culture and citizenship. The most extreme partisans of American interventionism exclaim that "affluence," liberals, Democrats, postmodernism and "white guilt" have made America soft to the point that the general population no longer has the courage to consent to the price of blood.[91]

This republican or traditionalist logic also sets them apart from the liberal supporters of the war who remind readers that military sacrifice today is a personal choice rather than a political obligation. These neo-liberals, or conservative liberals, defend a "voluntary" conception of sacrifice in line with a liberal ethic of individual action.[92]

> It seems there are some in the crowd who bemoan loudly that Americans are not sacrificing equitably in this war effort. My understanding of the concept of "sacrifice" implies a voluntary giving of oneself . . .. Perhaps the administration can provide a list of worthy organizations and some inspiring words of encouragement to Americans to give a sacrificial monetary gift to show their support.[93]

Beyond monetary donations and involvement in the "war of ideas," *The Wall Street Journal* notes the involvement of NGOs such as "the "Semper Fi fund, the Archdiocese (Catholic) of the military and the more well known USO[. They] are just a few of many privately funded groups helping the military serve the health, spiritual and entertainment needs of our soldiers and their families."[94] An earlier article recounts a multitude of similar individual initiatives, ordinary citizens and business leaders who have spearheaded the construction of a military hospital like the Center for the Intrepid, adjacent to the Brooke Army Medical Center in San Antonio, Texas. Articles laud other individuals whose donations have helped to establish programmes and networks providing sports, recreation and emotional support for soldiers and their families.[95] The point here is that articles in the *Journal*, *USA Today* and *Newsweek* stress the *individual choices* made by service members to enrol in the armed forces[96] and the advantages of a combat force composed of loyal, professional soldiers. This conceptualisation of sacrifice as an individual choice, however, also operates by counteracting the 'social diagnostic' of unequal enlistment patterns identified by opponents of the war or partisans of the draft. In this sense, pro-draft and anti-draft positions within the political field of debate over the war are intimately tied to homological positions, republican and liberal, in the field of ideological production. But supporters of the draft do not necessarily break along a Democratic/Republican line. Support for the draft is tied to political strategy, not necessarily political affiliation. However, this strategy of appealing to patriotism and the ideal of shared sacrifice is dictated by the dominant argumentative logic of the political field in the context of the war. This also means that the republican/liberal divide over sacrifice does not completely mirror the progressive/conservative divide between the news sources themselves. The *New York Times* and *Wall Street Journal* opposition sets the tone here between republicanism and liberalism, while *Newsweek* and *USA Today* both showcase a strong leaning to the republican vision of sacrifice without questioning the war itself. If *The Nation* opposes both the war *and* the republican conception of sacrifice, one way to look at this is to suggest that the choice is dictated by the logic which pushes it to distinguish itself from the *New York Times*, that publication closest to its positions in the journalistic field (at least amongst those presented here), and evidently the dominant progressive publication in the US field of journalism in terms of circulation.

The State Responds

The debates in newspapers over sacrifice in the War in Iraq reveal profound grief, sorrow, anger and contention over the human cost of the war in Iraq and how it is distributed across the fields of social classes, politics, citizenship and across geographical space. This discomfort is mirrored in three different ways. It manifests first as a return of the quarrel between a republican conception of military service as a civic duty and political obligation, and a liberal conception of voluntary sacrifice founded on individual preference. This marks the cultural or *moral* manifestation of the crisis in the fields of culture and ideological production. Second, following Hay's model of crisis, social actors sow narratives of abuse and victimhood, identifying rival social and political groups as those bearing responsibility for society's ills. Kreps shows[97] that as public opinion melted away in support of George W. Bush's grand strategy, it increasingly looked to political leaders who could offer an alternative to the failing war as I will show. These tensions also take shape through the denunciation of the political leaders, elites of American society and ordinary citizens who shy away from military service at a time when their country would need them, all the while deploring the unfair burden borne by soldiers and military families. As I have noted earlier, this republican denunciation with both supporters and opponents of the war decrying the inequality of sacrifice that has been asked of Americans, hailing from disfavoured class fractions, echoes across the political spectrum; it is not limited to Democrats or opponents of the war. Nevertheless, the analysis of republican critics of the volunteer army is vigorously rejected by those apologists of liberalism and volunteerism who refuse a social diagnostic of military service and sacrifice. Proponents of the war go so far as to suggest that opponents of the war in Iraq are in league with the enemies of America outside of its borders. As David Campbell has argued, "liminal populations" within the American imperial society are identified to foreigners and denied the moral benefit of citizenship. This shows once more that the logics of imperial domination and citizenship do not flow from simple residence within the borders of the US or even nominal citizenship. Some supporters of the war at least associate citizenship to a moral and cultural community which supports the imperial society's logic of geopolitical domination over Iraq. A similar logic was at play when partisans of the war suggested that cultural transformations in America, white guilt and the decline of traditional values have weakened the moral fibre of citizenship and thus the imperial society's resolve. Again in this case, the success of the imperial society's geopolitical logic of domination seemed to presuppose that its citizens possessed a certain ethos of sacrifice and faith in the superiority of the American model of society. Thus if debates over sacrifice are profoundly polarising the political field in these two fundamental senses they are also polarising the field of citizenship and exposing tensions between the first- and second-class citizens of the imperial society.

Third, concern over recruitment and sacrifice dominated not only newspaper articles and editorials, but think-tank papers,[98] congressional reports, findings by the Government Accountability Office[99] and academic analyses. The 2008 Congressional Commission on the National Guards and the Reserves, the GOA and the Center for American Progress all paint the alarming portrait of a worn army, exhausted troops, the melting away of the US's operational military reserve,[100] depleted equipment, bleak long-term recruitment prospects both in the enlisted and lower officer ranks,[101] and over-stretched reserve forces. As Moskos wrote: "Mobilization on the scale needed for Iraq (and, to a lesser degree, Afghanistan) reveals that relying on the reserves for such missions is not a long term option, both because of the unpredictable disruptions they cause in personal life and the increased likelihood of casualties."[102]

If the crisis of the American imperial society first manifested in the acute intellectual, cultural and political polarisation over the burden of military sacrifice we examined through the lens of newspaper articles, it also begins to appear in the pattern of institutional response to concern over the incapacity of the armed forces, especially the reserves, to prolong their mission. As Hall writes, "Conflicts between the fundamental class forces, which hitherto formed principally on the terrain of economic life and struggle, only gradually, at point of extreme conflict escalating to the level of the state, are now precipitated on the terrain of the state itself, where all the critical political bargains are struck."[103] That the imperial society is in crisis does not entail that the US faced imminent revolutionary transformation. It means that the unpopularity of the Iraq war, in the context of another conflict in Afghanistan, depleted manpower and equipment levels, and political grievance over the continuation of the occupation and the inequality of military sacrifice forced the DoD and political leaders to take action. A state or regime may successfully contain a 'crisis,' but this success will entail an institutional response of some kind, either meeting some of the demands which may be asked of it, or by forcing an authoritarian resolution. As hegemony weakens, in this case the liberal doxa of volunteerism, the state will attempt "to bring about by *fiat* what [can] no longer be won by consent."[104] This was the first form of response to the crisis of the American imperial society which manifested as an exhaustion of both citizens and soldiers' consent over continued and prolonged sacrifice.

If the American state preserved most of the American people from feeling more directly the effects of the war, by avoiding the return to a draft for instance, this has only been accomplished by lifting an extra levy from those who had already sacrificed. That posture marked the authoritarian response to the manpower crisis of the imperial society. As the wars in Iraq and Afghanistan eroded at the all-volunteer army,[105] the armed forces were forced to rely on Stop-Loss[106] orders to keep soldiers whose contracts were near expiration from retiring. In 2006, some 50,000 soldiers had already been served stop-loss orders since the beginning of the wars in Iraq and Afghanistan.[107] By 2009, the *New York Times* reported this number had risen to 120,000, effectively more than doubling the service members affected.[108] "Were it not for the Stop-Loss policy . . . [there] simply would not be enough personnel for the Army to complete its missions."[109] Congressional Research Services analyst Charles Henning gives a much higher figure: 185,000 soldiers were stop-lossed since September 11th, 2001.[110] Given the size of the total forces from the Army deployed to Iraq and Afghanistan, weighing around 120,000 soldiers, approximately one in ten soldiers fighting overseas at any one time has been involuntarily retained.[111] Significantly, "[an independent study commissioned by the Marine Corps] concluded that if the DOD policy of allowing only a single 12-month mobilization and no further involuntary remobilization continued, the Marine Corps would simply run out of units to mobilize."[112] This is not without counting the number of times certain units have redeployed to combat theatres. The Center for American Progress suggests as many as 84,000 soldiers have been redeployed to Iraq and Afghanistan on more than one tour.[113] Of these, 10,000 have returned to combat no less than *five times*.[114]

In 2008, the balance of forces shifted in the imperial society's political field. President Barack Obama was elected on a platform which notably entailed removing US troops out of Iraq "within 16 months," thus by the end of the "summer of 2010."[115] Of course, this strategy needs to be evaluated in light of a much wider (geo)political context in both international relations, Iraq, Afghanistan, Pakistan, and domestic politics in the USA – another topic in itself.[116] I do not presume that it is a direct response to the narrow question of unequal military sacrifice. Domestically, I do advance it is an answer to the much larger pattern of opposition and concern over the continuation, handling of the war, the elusive perspective of a short- or even mid-term political victory, the treatment of veterans and service members, and a tarnished international image. Domestically, Obama's plat- form sought to campaign on his early opposition to the invasion of Iraq in 2002 all the while appearing as a leader who would not shy away from the use of military force.[117] The defeat of 2004 Democratic presidential can- didate John Kerry illustrated the danger candidates risked if they could be made to appear 'weak' or afraid of military engagement, and thus the impor- tance of martial qualities for any presidential candidate in the political field. Though this had long been an integral part of Obama's grand strategy, the logic of the political field dictated for the candidate to distinguish himself from his adversaries, rival Democrat Hillary Clinton and Republican can- didate John McCain and further determined Obama's Afghanistan-not-Iraq strategy. One must also consider the fact that Obama inherited two wars from the previous administration, a legacy which also constrained the strate- gic options which confronted a diminished military capacity and a wide variety of possible commitments: concerns over China, Iran, North Korea, and of course, Afghanistan and Pakistan.[118] In this sense, the strategy of disengagement from Iraq is the crown jewel amongst many institutional and political responses that the American state (electoral politics are still an extremely regulated state process) has affected to contain the crisis posed by the cumulative effects of dwindling public opinion on the conflict in Iraq even as the public rediscovered the "Forgotten War"[119] in Afghanistan, torture scandals, and the erosion of the all-volunteer army. The phased with- drawal of American troops from Iraq is not evidence that the crisis was not serious or decisive. President Nixon slowly pulled out troops from Vietnam between 1973 and 1975, yet the outcome of the conflict as a decisive political defeat, domestically and internationally, and the notion that both the presi- dent and the war faced a crisis of legitimacy domestically is unquestioned. Democrats, like most opinion makers, hesitantly conceded that the troop 'surge' in 2007[120] was a success and that the military situation in Iraq was stabilising. Like Vietnam, Iraq is not a military defeat in the sense of conven- tional warfare. The US does not run an imminent risk of having most of its units overrun or destroyed in the short term by adverse forces as did the US army at Pusan in Korea (1950), the French Expeditionary Corps at Dien Bien Phû (1954), or the Allies at Dunkirk (1940). Like the "Peace with honour" strategy embraced by Nixon, phased withdrawal from Iraq only illustrates that Obama has accepted to forego a decisive or even partial victory in Iraq. But he cannot risk seeming politically weak by 'cutting and running' and advocating immediate retreat which, strategically, would arguably compro- mise the political and military gains of the surge, or at least those *alleged* to have been made by the White House and in the eyes of the media,[121] and compromise his image as a leader capable of employing military force if necessary. The point is that the War in Iraq faced a sufficient crisis of legitimacy that a *black* presidential candidate could be elected on a platform which notably involved ending the conflict.

As protests over the War aims in Iraq and the literal abuse of the soldiers' patriotism polarised the political field, the state responded by deciding to phase out and ending stop-loss by 2011, as announced by the DoD on 18 March 2009.[122] Like the decision to withdraw from Iraq, this corresponds to a second phase of the crisis in which the state begins to make concessions and transform its personnel policies to meet its manpower requirements. In order to achieve this, the Defense Department relaxed its recruitment standards to accept low-scoring applicants, high school dropouts and ex-cons.[123] Furthermore, as of 2006, nearly 40,000 foreign nationals were fighting in the US armed forces[124] hoping for a fast track to citizenship through the Armed Forces Naturalization Act of 2003 (H.R. 1954). According to the *New York Times*, after residing in America for two or more years, skilled immigrants in certain categories who enlist in the armed forces will be eligible for citizenship "in as little as six months."[125] Only recently, the Chairman of the Joint Chiefs of Staff publicly supported ending the "don't ask don't tell policy" and the ban on open homosexuality in the armed forces initiated under Bill Clinton. "No matter how I look at the issue, I cannot escape being troubled by the fact that we have in place a policy which forces young men and women to lie about who they are in order to defend their fellow citizens."[126]

In order to successfully meet the manpower requirements necessary for its place of geopolitical domination in international society, paradoxically, the US imperial society had to relax the borders that guard access to that most sacred of places: the field of citizenship from those formerly undesirable populations, both migrants and homosexuals. Finally, the Gates Defense Department obtained from Congress an increase in the Army's authorised end-strength by nearly 65,000 soldiers to 547,000 soldiers in the Active Duty component.[127] Aided by the economic crisis,[128] in some cases, the services doubled the enlistment and reenlistment bonuses they offered potential recruits in order to meet their targets for fiscal years 2006 through 2008.[129] Thus the state has met the crisis provoked by the twin wars in Afghanistan and Iraq by directly mediating social relations between the classes and expanding sacrifice upward the social pyramid through larger financial incentives, by promising to reduce the strain on those already serving, and ultimately putting an end to American participation in the War in Iraq.

CONCLUSION

Whereas traditional debates over empire define imperial power as economic or political rule over 'foreign' populations or territories, the imperial society framework highlights the fact that foreignness and thus imperial relations are first and foremost a question of citizenship, only partially determined by one's geographical position within or without the state's territorial borders. Rather citizenship is a continuum of exclusion and inclusion which can be measured against one's possession of different quantities of cultural and economic capital and the real life-opportunities but also obligations that these instantiate. Imperial domination like citizenship cannot be simply reduced to a self/other binary but rather should be conceived of as a graded set of rights, opportunities and obligations in a field of struggle. In this sense, the foreign is delimited by the outer-border of the field of citizenship, a shifting cultural and political construct which is only partially tied to geography. This said, geography *does remain* fundamental as one of the most powerful political markers which constructs and polices belonging to a particular community and produces in doing so the collective understanding of citizenship. It is fundamental in the second sense that one's geographical position can favour or hinder the acquisition and possession of cultural and economic capital and thus lead to radically different opportunities and life chances: in this case a life of sacrifice or one of shopping.

The imperial society and field of citizenship framework help us further highlight the ambiguous nature of the term 'foreign' through the dual nature of the crisis the US has undergone as a result of the war in Iraq. It is both a geopolitical crisis of a dominant society[130] and a crisis of the relationship between first-class citizenship and soldiering within the borders of this society. Inquiring about what looking at war beyond the battlefield entails when it comes to the American imperial society context as it still faces the wars in Iraq and Afghanistan necessitates a framework that underscores the impact of geopolitical competition and military conflict on the body politic. Once a hallmark of first-class citizenship, contemporary social and political mutations suggest that the position of soldiering has fallen toward the periphery of the field of citizenship. Increasingly, 'full citizens' – the dominant class fractions with larger reserves of economic and cultural capital, or those who inhabit urban geographical locations favouring the acquisition of these resources – seem to be the ones spared the burden of war. First-class citizenship is no longer tied to sacrifice; on the contrary, it is now defined as economic and cultural *overqualification* for the duty of sacrifice. We then see that soldiers, and especially white male soldiers, find themselves enmeshed in a web of borderline-imperial exclusion. I say borderline-imperial exclusion because these soldiers enjoy some of the symbolic attributes of first-class citizenship traditionally defined: whiteness and military service. At the same time, even these white soldiers become second-class citizens in the field of citizenship when we consider the disproportionate political obligation they must fulfil towards the state because of the specific habitus that their precarious reserves in cultural, economic and citizenship capital traced out. But we cannot go so far as to say they are "othered" simply because they are second-class citizens. Building on the contributions of postcolonial scholarship, the field of citizenship and the imperial society framework help us identify this form of imperial exclusion as only one degree along a larger continuum which cannot adequately be captured in the self/other binary of traditional definitions.[131]

The cultural renegotiation of the link between imperial citizenship and sacrifice is intrinsically tied to a renegotiation of the imperial society's place in the sphere of global politics during the course of a war which challenges the collective identity constructs of America as an exceptional nation. The War in Iraq throws into upheavals the horizon of social and cultural reference points which indicate to citizens of the imperial society where 'they fit-in' in the larger global society. Debates over individualism and rampant consumerism, inequalities of political obligation and socioeconomic cleavages between combatants and non-combatants, divisions between republicans and liberals,[132] opponents and proponents of the War, over the 'correct' distribution of military sacrifice amongst American citizens: all of these disruptions are manifestations of the same crisis of the American imperial society. The crisis espouses the contours and specific logic of each field[133] so that even semi-political rants about teeming shopping malls, in fact, express displaced anxieties about the crisis.[134]

If the crisis did not threaten social upheaval or mass demonstrations as occurred during the Vietnam War, it was nevertheless serious in two important ways. Institutionally, the crisis was serious enough that the Defense Department intervened in civil society by mediating social relations between the class fractions. After using multiple involuntary mobilisations of the Reserve forces and Stop-Loss orders, DoD relaxed recruitment standards to induct those the military deemed the undesirables of society, and increased sign-up and reenlistment bonuses. It also directly intervened in the imperial society's field of citizenship. Migrants would receive fast-tracked citizenship, and the armed forces now seemed ready to let homosexuals serve openly in the army. By increasing the rewards in economic capital in exchange for military service the DoD also potentially expanded sacrifice upward the social pyramid through larger financial incentives and promise to reduce the strain on those already serving. Finally, the cumulative effects of the crisis brought by the War in Iraq were serious enough that they escalated to the political level in such a way that the American public elected the first black presidential candidate on a platform which, economic considerations aside, promised disengagement from Iraq. This is not to suggest that Obama was elected *because* he opposed the war. Early opposition to the war however distinguished him from his two primary opponents Senators John McCain and Hillary Clinton and had been instrumental in gaining him great notoriety with opponents of the war.

The imperial society model can contribute to future inquiry by exploring in newspapers and the testimonies of soldiers and their families how the war in Iraq has generated collective experiences of moral anxiety about

America's place in the world as an exceptional nation. Future contributions on the question of sacrifice can explore the suspicious resemblance between private military violence and economically motivated recruits of the regular forces. This could help displace the problématique of private military service from the point of view of the logics which push states to employ armies or commercial violence workers to one which asks what are the implications of this decision for the politics of citizenship. This article and the imperial society framework already suggest that we examine these questions along a continuum of rights, obligations and capital reserves by focusing on the demographic composition of each population type (private and enlisted) as well as the symbolic rewards they hope to gain from service, as opposed to the narrowly material gains. Another avenue of inquiry can deepen the questions developed here about what fast-tracked citizenship for migrants and open service for homosexuals might suggest about the changing politics of citizenship in the imperial society. Furthermore, that the economic crisis of late 2008 helped the armed forces meet their recruitment objectives supports Deborah Cowen's proposition that military service has become a form of "workfare" for disenfranchised citizens over the last few decades.[135] This line of questioning can help one tie in an account of the crisis of the imperial society and the relationship between citizenship and soldiering into a larger account of a crisis of neo-liberal accumulation. The imperial society framework thus presents numerous possibilities to create dialogue between some of the most pressing questions facing international relations, critical geopolitics, cultural studies, political economy and classical theoretical concerns about citizenship, rights and political obligations.

NOTES

1. C. Moskos, 'A New Concept of the Citizen-Soldier', *Orbis* (Fall 2005) p. 669.

2. Amy Kaplan presents an excellent summary of the resurgent debate on American empire as well as a probing critique o the new apologists of empire. A. Kaplan, 'Violent Belongings and the Question of Empire Today. Presidential Address to the American Studies Association', *American Quarterly* 56/1 (March 2004) pp. 1–18. Laura Anne Stoler is interesting as well: L. A. Stoler, 'On Degrees of Imperial Sovereignty', *Public Culture* 18/1 (2006) pp. 125–146. For a sampler of viewpoints on the 'empire' debate in International Relations see A. Colas, *Empire* (London: Polity 2007); D. Grondin, 'Introduction: Coming to Terms with America's Liberal Hegemony/Empire', in D. Grondin and C.-P. David (eds.), *Hegemony or Empire? The Redefinition of US Power Under George W. Bush* (Aldershot: Ashgate 2006) pp. 1–17; J. N. Pieterse, *Globalization or Empire* (New York and London: Routledge 2004). For the key works of Americanists and historians on empire: W. Lafeber, *The New Empire. An Interpretation of American Expansion 1860–1898* (Ithaca and London: Cornell University Press 1963); W. A. Williams, *The Tragedy of American Diplomacy* (New York: Dell Publishing co. 1962); A. Kaplan and D. Pease (eds.), *Cultures of US Imperialism* (Durham and London: Duke University Press 1993); E. T. Love, *Race Over Empire. Racism and U.S. Imperialism 1865–1900* (Chapel Hill and London: University of North Carolina Press 2004).

3. K. Beitel, 'The US, Iraq and the Future of Empire', *Historical Materialism* 13/3 (2005) pp. 163–192; G. Arrighi, 'Hegemony Unravelling: I and II', *New Left Review* 32 (2006) pp. 23–80; Pieterse, *Globalization or Empire* (note 2) Chapter 4 in particular.

4. Though in Bourdieu's stead (see below), I do not subscribe to a 'unitary' view of the state, given the scope of this paper I only make a minimal effort to distinguish the bureaucracy, different institutions

and political actors at play in this process. See P. Bourdieu, 'Esprits d'États. Genèse et structure du champ bureaucratique', *Actes de la recherche en sciences sociales* 96/1 (1993) pp. 49–62.

5. See C. Charle, *La Crise des Sociétés Impériales. Allemagne, France, Grande-Bretagne 1900–1940* (Paris : Éditions du Seuil 2001); also C. Charle, 'Les "Sociétés Impériales" D'hier à Aujourd'hui. Quelques Propositions Pour Repenser L'Histoire du Second XXe Siècle en Europe', *Journal of Modern European History* 2 (2005) pp. 123–139; and C. Charle, 'Les Sociétés Impériales et la Mémoire de la Guerre: Allemagne, France, Grande-Bretagne', in F. Guedj et al. (eds.), *Le XXe Siècle des Guerres. Modernité et Barbarie* (Paris: La Découverte 2004) pp. 303–319 and 548–555.

6. P. Bourdieu, *La Distinction, Critique Sociale du Jugement* (Paris: Éditions de Minuit 1979); P. Bourdieu, *L'ontologie politique de Martin Heidegger* (Paris: Éditions de Minuit 1988). The use of Pierre Bourdieu in International Relations is only recent. Indeed, the sociological thought of Pierre Bourdieu has slowly been introduced by several key works in international political sociology, notably with Michael Williams, *Culture and Security: Symbolic Power and the Politics of International Security* (London and New York: Routledge 2007); and Vincent Pouliot, *International Security in Practice: The Politics of Nato-Russia Diplomacy* (Cambridge: Cambridge University Press 2010).

7. Charle, *Crise des Sociétés Impériales* (note 5).

8. Charle, 'Les "Sociétés Impériales" D'hier à Aujourd'hui' (note 5).

9. Ibid., p. 125 (author's translation). For major works on American exceptionalism and Manifest Destiny see A. Stephanson, *Manifest Destiny: American Expansion and the Empire of Right* (New York: Hill and Wang 1995); Kaplan, 'Violent Belongings' (note 2); M. Hunt, *Ideology and Foreign Policy* (New Haven and London: Yale University Press 1987). R. Horseman, *Race and Manifest Destiny. The Origins of American Racial Anglo-Saxonism* (Cambridge, MA: Cambridge University Press: 1983); M. McAllister, *Epic Encounters, Culture, Media and U.S. Interests in the Middle East since 1945*, updated ed. (Berkeley and Los Angeles: UC Press 2001).

10 See Chapter 1 in D. Cowen and E. Gilbert (eds.), *War, Citizenship, Territory* (New York: Routledge 2008); J. Agnew, 'The Territorial Trap: The Geographical Assumptions of International Relations Theory', *Review of International Political Economy* 1/1 (1994) pp. 53–80; Walker presents the classical and now canonical version of this argument: R. Walker, *Inside / Outside: International Relations as Political Theory* (Cambridge: Cambridge University Press 1993).

11. Typically hailed as a realist scholar of IR, Aron presents a thought-provoking problematisation of what 'sovereignty' actually means in the traditional definitions of 'empire' and 'imperialism' which precedes by decades the postcolonial critiques so common today. R. Aron, *République impériale. Les États-Unis dans le monde, 1945–1972* (Paris: Calmann-Lévy 1973); D. Campbell, *Writing Security. Unites States Foreign Policy and the Politics of Identity*, rev. ed. (Minneapolis: University of Minnesota Press 1998); Kaplan, 'Violent Belongings' (note 2).

12. Amy Kaplan makes this point very well. The post-911 nationalist narrative of the "homeland" evokes nativist tropes in American culture: a differentiation between real Americans of European ancestry and others who simply inhabit the territory. See Kaplan 'Violent Belongings' (note 2) p. 8.

13. D. Campbell, 'Global Inscription: How Foreign Policy Constitutes the United States', *Alternatives* 15 (1990) p. 275; Kaplan, 'Violent Belongings' (note 2).

14. A. Kaplan, 'Homeland Insecurities. Reflections on Language and Space', *Radical History Review* 85 (Winter 2003) pp. 86–87 (emphasis added).

15. A. Kaplan, *The Anarchy of Empire in the Making of US Culture* (Cambridge, MA, and London: Harvard University Press 2003).

16. Counsel of Economic Advisors to the President's Initiative on Race, *Changing America. Indicators of Social and Economic Well-Being By Race And Hispanic Origin* (Sep. 1998), available at <http://www.access.gpo.gov/eop/ca/index.html>, accessed 23 March 2008.

17. See Kaplan, 'Violent Belongings' (note 2); McAllister (note 9); and Campbell, *Writing Security* (note 11).

18. Kaplan, 'Homeland Insecurities' (note 14) p. 90 (emphasis added).

19. J. Butler, *Gender Trouble* (New York and London: Routledge 1990, 1999) p. x; E. Said, *L'Orientalisme: l'Orient Crée par l'Occident* (Paris: Éditions du Seuil 1980).

20. S. Dalby, 'Geopolitics and Global Security. Culture, Identity, and the POGO Syndrome', in S. Dalby and G. Ó Tuathail (eds.), *Rethinking Geopolitics* (London and New York: Routledge 1998) pp. 295–313; G. Ó Tuathail, *Critical Geopolitics* (Minneapolis: University of Minnesota Press 1996).

21. Campbell, *Writing Security* (note 11).

22. A. L. Stoler, 'On Degrees of Imperial Sovereignty' (note 2) p. 145 (emphasis added).

23. D. Campbell, 'Global Inscription' (note 13) p. 275.

24. Ibid., p. 277 (emphasis added).

25. L. Kerber, 'The Meanings of Citizenship', *The Journal of American History* 84/3 (Dec. 1997).

26. Bourdieu, *La Distinction* (note 6) p. 113.

27. Ibid., p. 128 (author's translation).

28. Note that for Bourdieu class compounds all of these possible vectors of dominance, including age and geographic location. Ibid.

29. See C. Snyder, 'The Citizen-Soldier Tradition and Gender Integration of the US Military', *Armed Forces & Society* 29/2 (Winter 2003) p. 187.

30. Inversely, however, if military service constitutes a civic obligation, it is often tied to benefits enjoyed in the community and often economic advantages.

31. On the citizen-soldier tradition in America and military sociology in general see Moskos, 'A New Concept of the Citizen-Soldier' (note 1); also C. Moskos, 'Toward a Postmodern Military: The United States as a Paradigm', in C. Moskos, J. A. Williams, and D. Segal (eds.), *The Postmodern Military* (New York: Oxford University Press 2000) pp. 32–50; C. Moskos with J. S. Butler, *All That We can Be. Leadership and Racial Integration the Army Way* (New York: Basic Books 1997); R. Krebs, *Fighting for Rights: Military Service and the Politics of Citizenship* (Ithaca, NY: Cornell University Press 2006); Brian Gifford, 'Combat Casualties and Race: What Can We Learn from the 2003–2004 Iraq Conflict', *Armed Forces & Society* 2 (Winter 2005) pp. 201–225; D. Segal and M. Segal, 'America's Military Population', *Population Bulletin* 59/4 (Dec. 2004); I. Feinman, *Citizenship Rites. Feminist Soldiers & Feminist Antimilitarists* (New York: New York University Press 2000); R. F. Titunik, 'The First Wave: Gender Integration and Military Culture', *Armed Forces & Society* 26/2 (2000) pp. 229–257. On the National Guards and Reserves read 'Appendix 5: History of the Reserve Forces', in the *Commission on the National Guard And Reserves* (31 Jan. 2008), available at <http://www.bespacific.com/mt/archives/017340.html>, accessed 8 Sep. 2009; M. Morgan, 'Army Recruiting and the Civil Military-Gap', *Parameters: US Army War College* 31/2 (Summer 2001) p. 101; D. Segal and N. Verdugo, 'Demographic Trends and Personnel Policies as Determinants of the Racial Composition of the Army', *Armed Forces and Society* 20/4 (Summer 1994) pp. 619–632.

32. For a fairly detailed list of these exclusions see Chapter 1 in R. Smith, *Civic Ideals: Conflicting Visions of Citizenship in U.S. History* (New Haven: Yale University Press 1997); also Kerber, 'The Meanings of Citizenship' (note 25); on Gays and Lesbians see Feinman (note 31).

33. M. Sandel, *Democracy's Discontent. America in Search of a Public Philosophy* (Cambridge, MA: Harvard University Press 1996, 1998) p. 318.

34. Krebs (note 31).

35. On the struggles of Women and gays to integrate the army and the ties between gender roles and military service see Feinman (note 31); Titunik (note 31); Snyder (note 29). Also see Krebs (note 31) pp. 167–171, for a short but effective summary of the challenges and struggles of Japanese Americans to obtain full citizenship rights.

36. Feinman (note 31) p. 135 (italics in original text).

37. Segal and Verdugo (note 31).

38. D. Cowen, 'Fighting for "Freedom": The End of Conscription in the United States and the Neoliberal Project of Citizenship', *Citizenship Studies* 10/2 (2006) p. 171.

39. See Moskos, 'A New Concept of the Citizen-Soldier' (note 1); Feinman (note 31). Also the dramatic fall over the decades following the Vietnam War in members of the House of Representatives who are veterans would tend to reinforce this hypothesis. M. Morgan (note 31) p. 101.

40. As of 2009, Active Duty forces (enlisted and officers) numbered 1.4 million men and women, while as of 2007 the Ready Reserve counted 1,088,587 service members (excluding members of the Standby and Retired Reserves). See Defense Manpower Data Center, *Armed Forces Strength Figures for July 31, 2009*, available at <http://siadapp.dmdc.osd.mil/personnel/MILITARY/ms0.pdf>, accessed 6 Sep. 2009; and Congressional Research Service Report for Congress, *Reserve Personnel Issues: Questions and Answers* (14 March 2008) p. 4 for data on the Reserves. Compared to a population of 304 million citizens as of the 2008 estimates; US Census Bureau, <http://quickfacts.census.gov/qfd/states/00000.html>, accessed 8 Sep. 2009.

41. D. Cowen, 'National Soldiers and the War on Cities', *Theory, Culture and Society* 10/2 (2007) para. 28.

42. T. Kane, *Who Bears the Burden? Demographic Characteristics of U.S. Military Recruits Before and After 9/11* (Washington, DC: The Heritage Foundation Center for Data Analysis 7 Nov. 2005) p. 1.

43. J. G. Bachman, D. Segal, P. F. Doan, and P. M. O'Malley, 'Who Chooses Military Service? Correlates of Propensity and Enlistment in the U.S. Armed Forces', *Military Psychology* 12/1 (2000) pp. 1–30; E. F. Tripp (ed.), *Surviving Iraq. Soldiers' Stories* (Northampton, MA: Olive Branch Press 2008).

44. L. Beeghley, *The Structure of Social Stratification in the United States* (Boston: Alyn and Bacon 1989).

45. Ibid.

46. Bourdieu, *La Distinction* (note 6) p. 123.

47. Ibid., p. 195.

48. See Segal and Segal (note 31) p. 23. Also see Gifford (note 31).

49. Moskos, 'A New Concept of the Citizen-Soldier' (note 1).

50. "Homologies" define the correspondences between dominant and dominated positions from one field to the next, i.e., economic, political, cultural, the field of citizenship; see Bourdieu, *La Distinction* (note 6).

51. Tripp (note 43) p. xviii.

52. Cowen, 'National Soldiers' (note 41) p. 28.

53. Ibid., p. 32.

54. B. Bishop and W. O'Hare, *U.S. Rural Soldiers Account for a Disproportionately High Share of Casualties in Iraq and Afghanistan* (Durham, NH: Carsey Institute Fall 2006) p. 1, available at <64.233.169.104/search?q=cache:b9iuIK0Yv3wJ:www.carseyinstitute.unh.edu/documents/RuralDead_fact_revised.pdf+carsey+institute+o%27hare&hl=en&ct=clnk&cd=1&gl=us&client=opera>, accessed 21 Oct. 2008.

55. L. Wacquant, 'The Social Logic of Boxing in Black Chicago. Toward a Sociology of Pugilism', *Sociology of Sport Journal* 9 (1992) pp. 221–254; Beeghley (note 44).

56. Bourdieu, *La Distinction* (note 6).

57. Segal and Segal (note 31).

58. Cowen, 'National Soldiers' (note 41).

59. See Segal and Verdugo (note 31) for two different perspectives on the question.

60. Ibid.

61. Ibid.

62. H. Fischer, 'United States Military Casualty Statistics: Operation Iraqi Freedom and Operation Enduring Freedom', *Congressional Research Services* (2008) p. 1.

63. I draw on three main inspirations here: the definitions of 'crisis' given by C. Hay, 'Narrating Crisis: The Discursive Construction of the 'Winter of Discontent', *Sociology* 30/2 pp. 253–277; S. Hall, *Policing the Crisis. Mugging, the State, and Law and Order* (London: MacMillan Press 1978); Bourdieu, *La Distinction* (note 6).

64. Ibid.

65. Hay (note 63).

66. For one, the journalistic analysis is meant to illustrate the tensions over military sacrifice in American society rather than reconstruct a thorough portrait of the media's coverage of the war. Such an undertaking can be found in other studies. It is not the object of this paper. See Bennett et al. which is particularly well-researched and thought-provoking: D. Kellner, 'The Media and the Crisis of Democracy in the Age of Bush-2', *Communication and Critical/Cultural Studies* 1/1 (March 2004) pp. 29–58; T. Groseclose and J. Milyo, 'A Measure of Media Bias', *The Quarterly Journal of Economics* CXX/4 (Nov. 2005) pp. 1191–1237; W. L. Bennett, R. G. Lawrence, and S. Livingstone, *When the Press Fails. Political Power and the News Media from Iraq to Katrina* (Chicago: The University of Chicago Press 2007); R. Entmann, S. Livingstone, and J. Kim, 'Doomed to Repeat: Iraq News 2002–2007', *American Behavioral Scientist* 52/5 (Jan. 2009); A. G. Nikolaev and D. V. Porpora, 'Talking War: How Elite Newspaper Editorials and Opinion Pieces Debated the Attack on Iraq', *Sociological Focus* (Feb. 2007).

I generated the documentary mass of this study with search strings in the Gale Academic Onefile Database and ProQuest and searched these articles using the keywords "war" and "sacrifice", which yielded both a manageable and pertinent amount of documentation. The three newspapers ranked highest in circulation in the United States. As for the news magazines, I selected them based on a practical and theoretical understanding of the journalistic field in the United States. *Newsweek*, publishing positions similar to *USA Today* in political tone, ranks at number 21 amongst the top twenty-five magazines, not necessarily news, while *The Nation* – the most vocal and antiwar of all publications – fails to register among the top magazines. The data was retrieved from Burrelles*Luce*, a self-described

"PR Professional": *2009 Top Media Outlets: Newspapers, Blogs, Consumer Magazines & Social Networks*, available at <http://www.burrellesluce.com/system/files/Top100Sheet6.24.09.pdf>, accessed 17 May 2010.

Though the string search only returned four articles for *The Nation* based on queries in the "abstract" or "keyword" section, I do not believe that a larger search would fundamentally alter the tone or political orientation of the articles. Keeping with the bourdieusian framework, I chose to recreate a portrait of the journalistic field based on the working assumption that the output of news or political views in any one paper was fundamentally relational to the positions espoused by other publications. Bourdieu argues that the "field of production" (newspaper output in this case) is partially structured by correspondences with the different positions in other fields (i.e., the fields of consumption, ideological production and, by extension, in the political field (liberal, conservative, etc). It is precisely because of this that Bennett et al., and Hall are critical of the presumed distinction between 'politically oriented' editorials, purportedly 'factual' news articles, and opinion letters which, ultimately, are also selected by the editorial boards. A critical political economy of news production and news rooms as organisations dictates against clearly distinguishing between the opinions of individual journalists, opinion letters, and the news organisations which employ or publish them.

67. There were 115 articles excluding the *Wall Street Journal*.

68. Kane (note 42) p. 1.

69. Bennett, Lawrence, and Livingstone (note 66) p. 79. See also Sarah Kreps, 'American Grand Strategy After Iraq' *Orbis* (Fall 2009) pp. 629–645.

70. See D. Denvir, 'Game Theory. In Northeast Philly, the U.S. Army is Waging a War for our Hearts and Minds', *Philadelphia City Paper*, 15 Jan. 2007, p. 8. J. Dao, '2,000 Dead: As Iraq Tours Stretch On, a Grim Mark', *The New York Times*, 26 Oct. 2005, available at <http://www.nytimes.com/2005/10/26/international/middleeast/26deaths.html>. By the late summer of 2007, the *Times* notes that black enlistment has dropped dramatically, from 20 percent "among active-duty recruits" in 2001 to 13 percent in 2006. S. Abruzzese, 'Iraq War Brings Drop in Black Enlistees', *The New York Times*, 22 Aug. 2007, available at <http://www.nytimes.com/2007/08/22/washington/22recruit.html?scp=1&sq=Iraq+war+brings+drop+in+black+enlistees.&st=yt>.

71. Denise M. Bostdorff, 'Judgment, Experience, and Leadership: Candidate Debates on the Iraq War in the 2008 Presidential Election', *Rhetoric & Public Affairs* 12/2 (Summer 2009) pp. 223–277.

72. Stephen J. Randall, 'The American Foreign Policy Transition: Barack Obama in Power', *Journal of Military and Strategic Studies* 11/1–2 (Fall and Winter 2008/2009) pp. 1–24.

73. See also A. R. Hunt, 'Lion's Roar, Lamb's Sacrifice', *The Wall Street Journal*, 3 June 2004, p. A15; and 'In Iraq, a Necessary Sacrifice? (Letter to the Editor)', *The New York Times*, 28 Aug. 2003, p. A30.

74. B. Herbert, 'Sharing The Sacrifice, or Ending It (Editorial Desk)', *The New York Times*, 8 Dec. 2005, p. A39(L); C. B. Rangel, 'Recruit with Patriotism, not Economic Incentives', *The Wall Street Journal* (Eastern Edition), 11 Oct. 2004, p. A19.

75. 'Faces of Christmas (Editorial)', *USA Today*, 22 Dec. 2006, p. 19A.

76. 'Troops, Families Know too Well the Sacrifice You Don't See (NEWS)(Editorial)', *USA Today*, 25 May 2007, p. 19A.

77. W. Bardenwerper, 'Party Here, Sacrifice Over There (Editorial Desk)', *The New York Times*, 20 Oct. 2007, p. A17(L).

78. Ibid.

79. R. Hampson, 'A Small Town Mourns its big Sacrifice in Iraq (NEWS)', *USA Today*, 25 Jan. 2008, p. 01A.

80. 'Troops,' (note 76).

81. K. Pollitt, 'Do You Feel a Draft', *The Nation*, 7 June 2004, p. 9; 'Desperate Heroics of World War II Don't Apply to Iraq (Letter to the Editor)', *The Wall Street Journal*, 5 May 2007, p. A.7.

82. P. Krugman, 'Sacrifice Is for Suckers (Editorial Desk)', *The New York Times*, 6 July 2007, p. A15(L); D. R. Gergen, 'A Nation in Search of Its Mission. What Can a Citizen do to Fight this War? (Column)', *The New York Times*, 17 Jun. 2002, pp. A21(N), A17(L).

83. 'Where is the Clarion Call to Arms? (Letter to the Editor)', *The New York Times*, 9 Mar. 2007 p. A22(L); P. Krugman, 'Other People's Sacrifice (Editorial Desk)', *The New York Times*, 9 Sep. 2003, p. A29.

84. B. Herbert, 'Consider the Living, (Editorial Desk)', *The New York Times*, 29 May 2006, p. A15(L).

85. Bourdieu, *La Distinction* (note 6).

86. Pollitt (note 81).

87. Ibid.

88. Tina Managhan's article in this special issue exposes well how the sheer impossibility of dissent in this national context to which antiwar geopolitics must adapt.

89. 'For Future Generations, Was it Worth the Sacrifice? (NEWS)(Letter to the Editor)', *USA Today*, 29 Jan. 2007, p. 12A (emphasis added).

90. M. Finelli, 'Why We Need a Draft. An Iraq Veteran and 9/11 Survivor Says That We Cannot Win Until All Americans Sacrifice', *Newsweek*, 10 Sep. 2007, p. 39.

91. Jeffrey Zaslow with contributions from E. Bernstein, 'Our Unceasing Ambivalence', *The Wall Street Journal* (Eastern Edition), 8 Dec. 2006, p. A.16; S. Steele, 'White Guilt and the Western Past', *The Wall Street Journal* (Eastern Edition), 2 May 2006, p. A16; 'For Future Generations' (note 89).

92. J. Dao, 'SUPPORT OUR TROOPS (Week in Review Desk)(2004: In a Word)', *The New York Times*, 26 Dec. 2004, p. WK5(L); D. M. Kennedy, 'What is Patriotism Without Sacrifice?', *The New York Times*, 16 Feb. 2003, pp. WK3(N), WK3(L); Gergen (note 82).

93. 'How to Fulfill a Desire to "Sacrifice" for the War (Letters to the Editor)', *The Wall Street Journal* (Eastern Edition), 21 May 2007, p. A15.

94. Ibid.

95. J. Gurwitz, 'Cross Country: Help for the Intrepid', *The Wall Street Journal* (Eastern Edition), 10 Feb. 2007) p. A8.

96. See for instance A. Goodnough, 'In War Debate, Parents of Fallen Children Are United Only in Grief. (National Desk)(United States)', *The New York Times*, 28 Aug. 2005, p. A1(L); T. T. Gygax, 'Sorrow and Debate', *Newsweek*, 15 Aug. 2005, p. 24.

97. Kreps (note 69).

98. L. Korb, P. Rundlet, M. Bergman, S. Duggan, and P. Juul, *Beyond the Call of Duty. A Comprehensive Review of the Overuse of the Army in the Administration's War of Choice in Iraq* (Washington, DC: The Center for American Progress updated Aug. 2007); L. Korb and S. Duggan, 'An All-Volunteer Army? Recruitment and its Problems', *PSonline* (July 2007), available at <www.apsa.net>; *Commission on the National Guard and Reserves* (note 31). Also Kane (note 42).

99. Government Accountability Office, *Reserve Forces: Army National Guard and Army Reserve Readiness for 21st Century Challenges. Testimony Before the Commission on the National Guard and Reserves* (21 Sep. 2006).

100. Kreps (note 69).

101. Ibid.

102. Moskos, 'A New Concept of the Citizen-Soldier' (note 1) p. 669.

103. Hall (note 63) pp. 318–319.

104. Ibid., p. 284.

105. *Commission on the National Guard and Reserves* (note 31).

106. See "5.1.8.4. Stop Loss. Under Section 12305 of reference (b), in DOD Directive 1235.12 (19 Jan. 1996) 'Accessing the Ready Reserves', p. 9; Department of Defense (24 Sep. 2001) 'DOD AUTHORIZES STOP-LOSS', New release from the Office of the Assistant Secretary of Defense (Public Affairs), available at <http://www.defenselink.mil/releases/release.aspx?releaseid=3061>.

107. Korb et al. (note 98).

108. 'Stop-Lossing Stop-Loss (Editorial)', *The New York Times*, 21 March 2009, available at <http://www.nytimes.com/2009/03/22/opinion/22sun2.html>.

109. Korb and Duggan (note 98) p. 469.

110. C. Henning, 'U.S. Military Stop Loss Program: Key Questions and Answers', *Congressional Research Service* (10 July 2009) p. 11, available at <http://www.fas.org/sgp/crs/natsec/R40121.pdf>, accessed 4 Sep. 2009.

111. Ibid., p. 14

112. *Commission on the National Guard and the Reserves* (note 31) p. 237.

113. Korb and Duggan (note 98) p. 6.

114. *Commission on the National Guard and the Reserves* (note 31) p. 82.

115. CNNPolitics.com, <http://www.cnn.com/ELECTION/2008/issues/issues.iraq.html>, accessed 6 Sep. 2009.

116. For a detailed discussion of these questions and American grand strategy in the context of the Obama presidency see: R. Haass, and M. Indyk, 'Beyond Iraq: A New U.S. Strategy for the Middle East', *Foreign Affairs* 88/1 (Jan. 2009) pp. 41–58; John Nagl and Brian N. Burton, 'Striking the Balance. The Way Forward in Iraq', *World Policy Journal* (Winter 2008–2009) pp. 15–22; James Goldgeier, 'Making a

Difference? Evaluating the Impact of President Barack Obama', *UNISCI Discussion Papers* 22 (Jan. 2010) pp. 116–129; Bostdorff (note 71); Randall (note 72); Kreps (note 69).

117. Bostdorff (note 71).

118. Kreps (note 69).

119. S. Ricchiardi, 'The Forgotten War', *American Journalism Review* (Aug./Sep. 2006), available at <http://www.ajr.org/Article.asp?id=4162>, accessed 6 Sep. 2009.

120. Bostdorff (note 71); Entmann, Livingstone, and Kim (note 66).

121. Ibid.

122. Department of Defense, 'End to Stop Loss Announced', New release from the Office of the Assistant Secretary of Defense (Public Affairs, 18 March 2009), available at <http://www.defenselink. mil/releases/release.aspx?releaseid=12564>, accessed 4 Sep. 2009.

123. Korb and Duggan (note 98) p. 468.

124. Cowen, 'Fighting for "Freedom' (note 38) p. 178.

125. J. Preston, 'US Military Will Offer Path to Citizenship', *The New York Times*, 14 Feb. 2009, available at <http://www.nytimes.com/2009/02/15/us/15immig.html?_r=2&hp=&pagewanted=all>.

126. Admiral Mike Mullen cited in E. Bumiller, 'Top Defense Officials Seek To End "Don't Ask Don't Tell', *The New York Times*, 2 Feb. 2010, available at <http://www.nytimes.com/2010/02/03/us/ politics/03military.html>, accessed 26 Feb. 2010.

127. Henning (note 110) p. 12.

128. L. Alvarez, 'More Americans Joining Military as Jobs Dwindle', *The New York Times*, 18 Jan. 2009, available at <http://www.nytimes.com/2009/01/19/us/19recruits.html?th=&emc=th& pagewanted=all>, accessed 19 Jan. 2009.

129. Henning (note 110) p. 10.

130. See references at note 3.

131. Such as the one Raymond Aron critiques; see Aron (note 11).

132. In the academic sense of these terms.

133. In these cases the fields of culture, citizenship, social classes, politics and ideological production respectively.

134. Hall (note 63).

135. Cowen, 'Fighting for "Freedom" (note 38).

Grieving Dead Soldiers, Disavowing Loss: Cindy Sheehan and The Im/possibility of the American Antiwar Movement

TINA MANAGHAN

Department of International Relations, Politics and Sociology, Oxford Brookes University, Oxford, UK

The article investigates the conditions of emergence of Cindy Sheehan (mother of soldier killed in Iraq) as a spokesperson of the American antiwar movement and its so-called 'spark.' It interrogates the emotional pull of the current 'support the troops' rhetoric and the usurpation of this and other patriotic signs and symbols by various antiwar groups as both a constraint on the realm of legitimate dissent and an enabling condition of intelligible subject formation – with particular attention given to the figure of the grieving mom. This article argues that the sympathetic, albeit tenuous, identification with this figure emerged through a simultaneous psychic identification with and disavowal of loss – with implications for the possibility and impossibility of dissent in the aftermath of 9/11.

INTRODUCTION

The article investigates the conditions of emergence of Cindy Sheehan (mother of soldier killed in Iraq) as a spokesperson of the American antiwar movement and its so-called 'spark.' It interrogates the emotional pull of the current 'support the troops' rhetoric and the usurpation of this and other patriotic signs and symbols by various antiwar groups as both a constraint on the realm of legitimate dissent and an enabling condition of intelligible subject formation – particularly the figure of the grieving 'mom.' This article argues that the sympathetic, albeit tenuous, identification with this figure emerged through a simultaneous psychic identification with and disavowal of loss, both of which aligned the war protester with the nation even in the course of dissent. Borrowing from Judith Butler, the imperative to support the troops may be described as a 'loss of the loss' such that the troops came to stand in for that which we could not bear to lose (and must vigorously protect and defend) and that which we already lost: amongst other things, an ideal of innocence in a nation that is dead and yet relentlessly persists.[1] The ability of Sheehan to coax latent anxieties into political action and speech – indeed, the intelligibility of Sheehan as a legitimate political actor – will be explored in terms of the gendered (and less obviously raced) socio-political narrative structures that shaped seemingly personal, affective responses to the current war in Iraq and in terms of the consequent im/possibility of dissent.

This article begins with the assumption that language 'acts' – i.e., that the subjects who speak it are also and always constituted by it. Contrary to suggestion that Sheehan's popularity resided in her ability to give voice to that which was 'already there, deep within her' – an alternative rationality or ethic based on authentic, maternal ways of knowing and being in the world – this paper will argue the inverse: that Sheehan herself became artic- ulated in and through language. The dominant discourses of motherhood, caring and attentive love, and the call to support the troops all preceded her. She became intelligible to us, and to herself, through language and, in very important ways, she did not exist outside of its structures. This is not to suggest, as have some of her critics, that she was merely a pawn – of 'the left,' 'of circumstance,' or even of language itself. Says Foucault, 'Discourse is not life; its time is not yours' which is, I think, to say that life exceeds language, but we do and we must live, work, and love within its bounds and constraints, even as we seek to alter these constraints and the terms of our existence.[2] Following Butler, I will suggest that the social identities to which we belong, ascribe, and resist are not of our own making and yet we fit and unfit ourselves into and out of them in order to 'be' at all, in order to speak, and, in order to undo those categories that we did not choose, yet hail us and seek to contain us.[3]

The task of genealogical analysis is to explore how subjectification pro- ceeds: how we become and un-become both in order to 'be' and to 'be other.' Cindy Sheehan said that she became a 'Peace Mom' the day her son died. She was arguably (re)constituted through both loss (as she no longer knew how to be who she 'was') and through language (the discursive realm of possibility). It is in this dual sense that we can begin to apprehend the psychic space of the subject and, *'the psychic life of power'* or the psychic life of dominant narratives, including the emotional pull of 'support the troops' rhetoric.[4] At stake in the narrative structures that shape our lives is not just our understanding of the world, but our status as subjects whose very viability depends on the sociolinguistic processes of recognition and naming. As Butler explains, 'Subjection exploits the desire for existence, where existence is always conferred from elsewhere; it marks a primary vulnerability to the Other in order to be.'[5] But, so long as it is the case that we are constituted in our relations to others and in the sociolinguistic structures of language, it is also necessarily the case that we can be 'undone' by the very relations and narrative structures that enable us to be *at all*.[6]

Loss, therefore, can take many forms. We can lose people, places, ideals, and cherished beliefs – all the necessary things that sustain us and our place in the world, informing who we are and where/how we fit in.[7] To persist in the face of loss we must re-become, re-attach, and/or disavow the losses that nonetheless form and compel us and our interactions with the world. Sheehan chose to re-become a 'Peace Mom,' a public figure, who, according to her own narrative, could stay silent no longer and rallied a grieving nation behind her. But, not all losses could be grieved by Sheehan, much less the nation, equally. Some losses were never spoken, could not be spoken, and could not be grieved or protested against.[8] These were the unavowable losses which became the formative grounds of that which was said and could be said – 'the limiting condition of [speech's] possibility'[9] – shaping the narrative structures within which Sheehan found her way again and became the public face of dissent. Sheehan's loss was political and the nation's loss was personal according to the culturally instituted, discursive forms of melancholia that governed the possibilities for intelligible subject formation in the face of both avowable and unavowable loss.

So, what were the spaces available for dissent within a melancholic social structure that demanded the recognition of some losses and the disavowal of others: such that Americans were compelled to grieve dead soldiers while disavowing other forms of loss and such that the compulsive command to speak one's support for the troops and their families demanded equally compulsive forms of forgetting and erasure? This paper will seek to answer these questions by exploring the emotional landscape the opposition to the war in Iraq relied upon as part of a geopolitics of patriotism and by tracing the narrative structures that have arisen in the aftermath of 9/11 to codify the nation's loss – specifically in terms of the subjects that have, in a sense, been written or spoken into, or out of, discursive existence. Cindy Sheehan, the grieving mother who became the 'face' of the antiwar movement and of the nation's loss, will figure prominently in my analysis, but so too will other subjects who came to figure prominently in national stories of loss and renewal: some by their incessant presence (such as the troops), some by their re-emergence (such as the firefighters who were re-imagined in terms of the 'Greatest Generation'), some by their marked and pervasive absence (including the ghosts of Vietnam), and some by their combined presence/absence (such as the Iraqi and the Vietnam Vet). It is in studying the complex interplay between presence and absence that this paper will begin to tease out how the imperative to support the troops functioned within the psychic life of the nation – giving rise to the figure of the troops that Americans were compelled to support, the figure of the 'grieving mom' as the voice of dissent, and Americans themselves as protesters, patriots, and/or otherwise conscientious citizen-subjects. This paper seeks to reassess the geopolitics of war abroad by exploring the animating tensions of war beyond the battlefield – in order to realise both how these have structured war itself and how war has invested a new more intimate space, the nation's psyche, bringing about a new geography of support and dissent for and from the war effort.

'SUPPORT THE TROOPS' AS A COMPULSORY UTTERANCE

'Support the troops, not the war' and 'Bring the troops home now' are two examples of the pro-troop slogans that were repeatedly employed by the antiwar protesters during the *active* phases of the Iraq War – so much so that they could be fairly described as the rallying cries of the mainstream American antiwar movement.[10] Indeed, Working Assets, a San Francisco-based long-distance firm and credit card company that supports 'liberal' causes, reported that of six different protest signs it offered on its website (for those against the Iraq War to download and attach to placards or bumper stickers), the one reading 'Support Our Troops, Bring Them Home Now' was the most popular and the one that the company chose to put up on its own billboards across the country. The president of Working Assets explained this choice as follows: 'We are trying to take back the language. Our message is that it's time for those who are antiwar to make it clear that we care just as much about our soldiers as those who are pro-war. . . . It really isn't some tactic.'[11] Similarly, according to the website 'bringthemhomenow.com,' by buying a thirty-nine-cent postage stamp featuring 'a yellow ribbon transposed over a peace sign' Americans can support veterans and citizen groups opposing the war while helping to 'spread the word that supporting our troops and fighting to bring them home now are one and the same.'[12] The idea that opposition to war naturally coincides with support for the troops merits some reflection in and of itself, but this paper is particularly intrigued by the social imperative underlying this relationship such that opposition to war *must* be seen to coincide with support for the troops.

Especially in the early days of the antiwar movement, expressions of support for the troops seemed to preface nearly any and all statements of opposition to war that were made before the mainstream press – and for good reason. As a journalist for the *San Francisco Chronicle* reported, Eddie Veder, the lead singer of the rock band Pearl Jam, was jeered and booed at a 2003 concert in Denver, Colorado when he criticised the war and the president on stage. It was not until he thought to add, 'Just to clarify . . . we support the troops,' that cheers and applause erupted from the audience again – enabling the show to go on.[13] This charged and emotional exchange, while seemingly spontaneous, followed what I will suggest was becoming an all too familiar script for anyone voicing dissent – in which supporting the troops became, more or less, a condition of legitimate dissent.

Of course, once a general consensus emerged that the United States could not 'win the war', this changed somewhat. The chorus of the war's detractors broadened and the pertinent questions shifted from whether or not you are *for* the war to *what went wrong?* and *who is to blame?* These latter questions opened new and retrospective bases for opposition to the war in Iraq, but they do not illuminate this paper's central concern with the original grounds on which opposition to the war in Iraq was articulated – i.e., when even strategic considerations were interpreted in terms of whether or not one was for or against the war effort and when 'being against' the war became inextricably entangled with 'being for' the troops.

THE FIGURE OF THE GRIEVING MOM

What interests me is not only that many protesters chose to preface their critique of the war with 'We Support the Troops,' but that Cindy Sheehan, the mother of a dead soldier, arguably ignited the mainstream antiwar movement with her emotional plea to 'support the troops' by bringing them home now. Sheehan was variously referred to as the 'catalyst,' the 'spark,' and even the 'Rosa Parks' of the American antiwar movement. Her story was featured in *Time Magazine*, The Today Show, the *New York Times*, CNN and a host of other national and local media outlets across the country. The fact that she and other grieving mothers and widowed wives were at the forefront of this antiwar movement in contrast to the young men who led the anti-war demonstrations to protest American intervention in Vietnam, did not go unnoticed.[14] Various reasons were given to account for the prominence of women, such as influence of the feminist movement on the current generation of women, the maternal responsibility to protect, and that which Celeste Zappala, a co-founder of Gold Star Families for Peace, described as 'a certain ferocity in motherhood.'[15] However, only the influence of feminism might be able to answer the question of 'why now?' And none of these explanations can explain why Sheehan was recognised by fellow Americans, male and female alike, as a suitable icon for the movement – one who came to symbolise 'the grieving mother in all of us.'[16] By what authority did she become the face of America's loss and the voice of the antiwar movement?

The death of twenty-four-year-old Casey Sheehan, who was killed in Iraq just five days after he arrived, precipitated Cindy Sheehan's own symbolic death of sorts – or at least her 'un-doing.' But, by her own account, it simultaneously precipitated her spiritual birth:

> I didn't know it then, but I know it now. When Casey died . . . he gave spiritual birth to his real mom. The real mom who was hiding behind her ignorance, faith, marriage, family, and comfort began to emerge on April 4. As I lay in a crumpled heap screaming . . . something snapped . . . I had to decide something in my heart and soul. Would I stay here and fall into a depression of grief and regret? Would I voluntarily leave and join Casey through suicide? Or would I stay and fight?[17]

As she described it, she not only chose to stay and fight, but was transformed from a private mom into a 'public peace mom' and a crusader who refused to be silent.[18] This in itself was arguably a performative act. Of the fact that she was transformed, I have no doubt. But the 'real' mom, I will suggest, only emerged in the process of re-becoming and in the telling of this discursive event.

There are at least two issues here. First, I am interested in the narrative resources that preceded and enabled this particular 'coming out', granting a grieving mom the authority to publicly voice the heretofore unspoken and, in the process, become the public face of dissent. Second, following Wendy Brown, I am interested in the 'compulsory' forms that her political speech and self narration have had to take and were condemned to follow as a condition of her intelligibility as a public person[19] – and the implications for both the possibility and impossibility of dissent. To the extent that we can focus on one and then the other, I will begin by focusing on the possibility of dissent that emerged in and through Sheehan's performative act – while keeping in mind Brown's critical questions about the *un-freedom* that inhered in this form of breaking silence. In the latter sections of the paper, I will focus on the disavowals that governed what could be said, seen and heard, marking the spaces of dissent's impossibility – while keeping in mind freedom's intransigence (as manifest in the failures to reiterate, re-do, repeat, and re-present) which mark the limitations of governance.[20] The two, in fact, are irrevocably intertwined.

As Butler reminds us, subjectification cannot precede without subjection. Sheehan's performative act may be explained, as per Butler, to the extent that her story echoed previous stories and 'accumulate[d] the force of authority through the repetition or citation of a prior and authoritative set of practices.'[21] Says Butler, 'The one who acts . . . acts precisely to the extent that he or she is constituted as an actor and, hence, operating within a linguistic field of enabling constraints from the outset.'[22] Linguistic agency emerges through the process of re-iteration or the process by which a subject re-iterates the discursive conditions of its own emergence in new ways or to different purpose or effect. In this case, the utterance 'I support the troops' was effectively usurped and re-staged by a seemingly familiar figure, the figure of the grieving mom. In so doing, its meaning changed and the discursive positionality of a 'soldier's mom' in the emotional economy of the nation was pluralised, such that the bourgeoning antiwar movement could claim that to support the troops and their families no longer meant to stifle one's misgivings about the war; it meant to vocalise them and demand the Bush administration send the troops home – 'NOW!'

Certainly, a movement was galvanised by Sheehan's performative act. But, as Terry Lovell explains, quoting Richard King, 'to protest publicly was itself to assume that one already belonged, that a space of public appearance waited to be rightly occupied.'[23] As various commentators suggested, Sheehan's question for the president, 'What noble cause [did my son die for]?' struck a chord at a time of growing public unease about the mounting casualties of American soldiers and about the purpose and direction of the war in Iraq. The sentiment expressed in much of the mainstream media was that she was 'tapping into a growing popular feeling that the Bush administration is out of touch with the realities, and the costs, of the Iraq war.'[24] It was a time when 56 percent of respondents in a CNN poll reported that they thought the war was going poorly.[25] It was also a time when, in the aftermath of President Bush's 2004 re-election, the opposition to the war was described as unfocused and seemed to have lost its momentum.[26] Although Sheehan was described by Tom Matzzie, the Washington director for MoveOn.org as 'a herald, waking everybody up,'[27] the evidence indicates that the American public was awaiting a call it could rally behind and Sheehan's poignant and personal message to the president was one they could identify as their own.

I think Lovell's analysis of the performative force of the actions of Rosa Parks – the black woman from Montgomery, Alabama who refused to sit at the back of the bus – is instructive in understanding the authority contained within Sheehan's act of resistance.[28] Lovell argues that the authority endowed on Parks's refusal did not derive from Parks alone. Building on and departing a bit from the work of Butler, Lovell emphasises that the authority bestowed on the act was a collective and interactive product of a movement-in-waiting and did not reside solely in Parks's ability to effectively re-stage a prior set of authoritative practices, conventionally reserved for white bodies. Parks was but one participant in what was a still fairly sporadic, but, at least for area bus drivers and passengers, a not completely unfamiliar act of resistance; in the months preceding two women were arrested for the very same act (and other instances were known to have occurred). But, Parks's act was bestowed with a retrospective authority through a 'double process of recognition' in which first local activists and then others (as judged in the political mobilisation that followed), took up her cause – deeming her 'a suitable candidate to be the standard bearer behind which the challenge to the bus laws would be mobilized.'[29] Parks, who is said to have embodied dignity and 'middle class respectability,' became an icon of the injustice suffered by the black community.[30] In recognising and assenting to the performative force and iconic status of her particular act, Lovell argues that the black community simultaneously constituted itself 'as well as the authority of its representatives.'[31] I believe Sheehan's authority resided in a similar process of double recognition.

There were pockets of resistance, moments of large-scale mobilisation, and, as already stated, a growing unease about the war in Iraq prior to the emergence of Cindy Sheehan on the national stage. What Sheehan did was re-energise the antiwar movement and enlarge the legitimate arena of public debate by articulating doubt and doing so with the legitimacy that could not be accorded to any leftist extremist, scruffy anarchist, college-age idealist, or an imagined 'troop-hating' Jane Fonda–type. Sheehan articulately and forcefully raised questions about the war and demanded answers from no less than the president himself, without apology and with a sense of entitlement that perhaps no one other than a soldier's mom could bring to bear on this protest act. Sheehan's decision in August of 2005 to set up a protest encampment on the outskirts of President Bush's ranch and her vow to remain there until the president, who was vacationing there at the time, came to speak with her and answer her question, 'What noble cause?', arguably became the largest national media event of summer's end. As an on-site CNN correspondent explained, 'She's a symbol with savvy, fully aware of the power of her own image, a mother who lost her son to war demanding a meeting with the man who sent him there.'[32]

Whether she could have been fully aware of the power of her own image is doubtful. In the beginning, there was just Sheehan and a handful of supporters. What is clear is that as a mother of a dead soldier, she felt entitled to usurp whatever authority was there to be had. Furthermore, it is apparent that various social justice organisations and the media were willing and eager to recognise her as a standard-bearer of the antiwar movement, endowing her actions (performatively) with *the authority that must be accorded to a grieving mother*. Thus while it has been said that Sheehan 'single-handedly reignited the antiwar movement,'[33] it is important to recognise that, like Parks, Sheehan's authority resided in a double process of recognition wherein the antiwar movement both designated her as its symbolic representative and was re-constituted through its relation to her. As a mother of a soldier killed in battle, Sheehan was deemed the authentic voice of national grievance and loss. In an attempt to explain her following, journalist Hillery Hugg put it this way:

> Many . . . seem drawn to what she's saying, reluctantly or surprisingly, moved by a connection to her primary grievance that this war was not worth the price – not worth the lives of other people's children, or for that matter, the lives of their own children – whether or not they agree with everything else she has to say about the broader political situation.[34]

The following excerpt from a newsletter written by the women's peace organisation, CodePink, suggests hers was a grievance that simply could not be denied: 'We are calling for everyone to join her in Crawford and stand with her . . . to bear witness to the deep loss of a mother who is demanding answers from the nation's President.'[35]

The power of Sheehan's message, I will suggest, may not have lain so much in what she said, but in her ability to say it at all – to give voice to the uncertainty and doubt that was already there, to speak without apology before a national audience, and, not least importantly, to command a certain respect even from those who adamantly opposed her. The figure of the grieving mom could, in fact, say and do that which nobody else could as it was at the behest of this figure, that others were silenced. In various municipal debates, appeals to hang yellow ribbons in support of the troops were often made by and/or in the name of soldier's families – particularly their 'moms.'[36] For example, a community debate that erupted in Camden, Maine, as a result of a municipal decision to remove yellow ribbons from lampposts (which were hung by the mother of a soldier deployed overseas), revolved around whether or not the yellow ribbons served as a political statement or 'a symbol of a mother's love for her child' – i.e., merely an 'innocent message of love and safe return.'[37] In Fieldsboro, New Jersey, the decision by council and the mayor to ban yellow ribbons from public property generated vociferous street protests. *WorldNetDaily* reported that 'one protester was spotted wearing a Saddam Hussein mask while carrying a sign that read: "Mayor Tyler is my buddy."'[38] Further, Diane Johnson, the woman who started the controversy by placing a 'dinner-plate-sized ribbon' on the town's official welcome sign, chided the mayor for his lack of sensitivity to the soldier's mothers: 'There are mothers in town who have sons over there. You think [the mayor] would be a little bit sensitive to them.'[39]

The traitorous identifications assigned to those who didn't openly and actively support the troops or the mothers left behind partly explains the speech that was afforded to war vets and mothers of soldiers – in particular Cindy Sheehan. Sheehan, moreover, was not just any mom, but, as described by one journalist, a particularly 'granola-y' type, wholesome and rather plain looking.[40] While known to be stubborn and strong-willed, the media also described her as generous with her hugs and affection.[41] As members of the media and the Bush Administration noted, her 'just-an-average-mom' persona was a powerful asset; there would be no 'swift-boating' of a Gold Star Mom. It is also worth bearing in mind precisely that which goes without saying – because a rather different tale and explanation would have emerged if it was otherwise: Sheehan is white. It is the unmarked (and un-remarked) character of her whiteness that enabled this particular grieving mother to occupy the space of and symbolically become the *grieving mother in all of 'us'* (a symbolic mother to the nation) – constituting a particular 'us' and nation in turn.[42] Defiance resided in this performative act, certainly, but the lack of audacity required should be remarked. In the words of Patricia Hill Collins, 'Current assumptions see African-Americans as having race, White women as having gender, [and] Black women as experiencing both race and gender . . . '[43] What this means, amongst other things, is that while the current president, Barack Obama, has had his very citizenship questioned, for Sheehan, the traitorous charges were primarily waged at the level of gender – a point to be discussed.

For now, the important point is that so long as the major media organisations and mainstream public recognised her as a symbolic mother to the nation and, hence, the legitimate voice of grievance and loss, she was a match for George Bush. As she camped outside his ranch for the duration of his vacation and the media picked up her cause, a national event was created – a stand-off between a mother and a president – such that, even if he ignored her, the president did not have the choice of *not* responding to her; it had become a matter of how.[44] A *New York Times* columnist noted that Sheehan was 'an opponent that had to be treated very gently even as she aggressively attacked the president and his policies.'[45] Similarly, *Time Magazine* acknowledged that 'Bush's team cannot fire back hard, as it usually does when criticized.'[46] So, the president treaded lightly, publicly expressing sympathy for her and respecting her right to protest.[47]

Bush responded indirectly to Sheehan and the corresponding media gaze that had fallen upon him in other ways as well. In addition to staged photo-ops of him hugging other mothers whose sons and husbands were deployed in Iraq, at the end of August in a speech to war veterans, the president notably departed from his previous practice of minimising the mounting death toll and recited the number of American soldiers killed in the war: 1,864.[48] This came from a president who was heavily criticised for censoring the costs of war (as measured in American lives) by barring the media from filming the return of caskets containing the nation's war dead.[49] What is, in fact, noteworthy about Bush's speech is first that the president had to respond to the personal demands of a grieving mom at all and second the extent to which the tables had turned, so to speak, in terms of who had the upper hand when it came to supporting the troops. Sheehan lambasted the president and his policies publicly and on a daily basis, seemingly speaking freely, while the president was on the defensive – forced to speak, albeit in throttled ways.

But, it would be wrong to take from this that Sheehan's speech was 'free' – even if rampant and largely uncensored (as judged by her internet blogs). Using Brown's analysis,[50] Sheehan's speech can be characterised by a certain 'compulsory discursivity.' What enabled her speech also constrained it. Her speech was effective to the extent that it was the speech of a grieving mom. As a result, the public identification afforded Sheehan was always somewhat tenuous. Interestingly, the media and the Bush administration knew this. In the words of *Time Magazine*, 'Once Sheehan starts acting like a politician, say some Republicans and even some Democrats, she will become just another voice in the debate – easy, in other words, to neutralize.'[51] To come to terms with her loss as the nation's became ever more difficult when she was made part of the national mediascape, which left her as a war dissenter.

Those intent on disparaging her, usually elements of the far right who took it upon themselves to defend president Bush and to say the things that he could not, usually did so by trying to paint her as either disingenuous or as consumed by emotion (and hence irrational) – and, awkwardly, sometimes as both. As indicated previously these attacks were highly gendered. She was, at times, charged with being a pawn of the Left because of her collaboration with Michael Moore and because some of her activities were financed by established antiwar organisations and advised by Fenton Communications, a public relations firm known to sponsor left-wing advocacy groups like MoveOn.org.[52] On his nationally syndicated radio show, Bill O'Reilly, for instance claimed that Sheehan was "being run by far-left elements who are using her, and she's dumb enough to allow it to happen."[53] Others similarly charged her with a dangerous emotionalism, calling her an 'emotional predator,' making unreasonable demands of a president who, as a 'wartime commander in chief' needs to retain a certain emotional distance from events in order to do his job effectively.[54] There were also those who portrayed her as intentionally manipulative – having left her husband and remaining (adult) children behind to pursue her own political agenda. Real grieving, it was suggested, was done quietly and in private; it was not a choreographed affair for public consumption.[55] This geography of support/dissent made her discursive subject space an impossible place to inhabit even though it undoubtedly offered her a powerful platform.

Thus, always haunting this sympathetic figure of the 'good' grieving mom or the 'peace mom' – the 'soft-spoken [and] selfless' mom 'who speaks only in soothing tones to serve the material and emotional needs of others' was the figure of the 'bad' mom – self-righteous, shrill, unreasonable, manipulative, and pursuing her own agenda to the detriment of those who needed her.[56] Against these charges, Sheehan aptly situated her actions within the discourse of maternal thinking:[57]

> I have received dozens of emails with this heading: Go Home and Take Care of Your Kids . . . I think of all the name calling and unnecessary and untrue trashing of my character, this one offends me the most . . . . First . . . because it is so blatantly sexist . . . second . . . is that I believe that what I am doing is for my children, and the world's children . . . I think that the strategy of eternal baseless war for corporate profit and greed is bad for all of our children: born and unborn . . . . Constant war is not a family value.[58]

This discourse of maternal rationality, combined with her sacred and unde-niable loss, granted her speech a certain authenticity and authority which afforded her a lot of leeway with the mainstream media and general public.

Even her potentially controversial comments to a CBS reporter in which she referred to the Iraqi insurgents as 'freedom fighters'[59] – was, much to the chagrin of the right-wing political commentator, Bill O'Reilly, not picked up in the mainstream press. In fact, when O'Reilly tried to raise this point on the *Late Show with David Letterman*, suggesting that Sheehan had stepped out of bounds (indeed) and should be more careful about what she says, Letterman (with spontaneous applause from the audience) retorted: 'Well, and you should be very careful with what you say also . . . . How can you possibly take exception with the motivation and the position of someone like Cindy Sheehan?'[60] Except for far right elements, many Americans expressed empathy for Sheehan and, regardless of whether or not they completely agreed with her, were willing to grant her opposition to the war a public hearing. She had become a symbol of a movement, the legitimate voice of dissent – so much so that perhaps what she symbolised was more important than what she said. As a result, the more politically controversial things she did say – i.e., things that did not simply correspond to the grieving mom image that the media itself helped to stylise – were, often times, simply unheard.

But, I will argue that this leeway was afforded on account of her align-ment with the troops – on account of the fact that her protest and vigil, was not waged *primarily* on behalf of the world's children, but on behalf of her son, a soldier.[61] This suggests that what was ultimately being protected was the figure of the soldier who had come to embody no less than the promise of the nation. By supporting the troops, dissenters could distance themselves from a cultural memory of betrayal associated with antiwar protesters and other naysayers from the Vietnam era – whom, as Linda Boose notes, were often recalled in the figure of a woman:

> In story after story that began in the 1980s to pour forth from Vietnam veterans, any and all rejection that some of them may have experi-enced upon return – together with guilt for actions in Vietnam that such stories disguise beneath the figure if a rejecting external accuser – was remembered in the person of a woman: a wife, a sister, a girl-friend, an airline stewardess, or even, in the 1988 film Hamburger Hill, an invented account of soldiers being greeted upon return by Berkeley coeds throwing dog shit.[62]

Metonymically, it was the feminised protester who crippled, emasculating men and an entire generation. Sheehan's most vehement critics, the counter-protesters who staged pro-war rallies outside of her encampment, recalled this gendered legacy, trying to recast Sheehan as a cold-hearted, traitorous bitch with protests signs which read, 'Bin Laden says keep up the good work Cindy,' and 'How to wreck your family in 30 days by "Bitch in the Ditch."'[63] These signs called to mind an image of not just a bad mom, but an emasculating one who would betray her husband, her son and nation.

It is for this reason significant that Sheehan waged her campaign with the strong backing of many war veterans. War vets had a significant presence at Camp Casey and toured with Sheehan during the cross-country Bring Them Home Now Tour which was launched on the final day of the Camp Casey Encampment (31 August 2005). In addition to the antiwar organisations developed by and on behalf of military families, including Sheehan's Gold Star Families for Peace and Military Families Speak Out, it was veterans groups like Operation Truth and Iraq Veterans against the War that were at the forefront of the antiwar movement.[64] Arguably much like Sheehan, the veterans felt a certain entitlement to their dissenting opinions – which had been denied others on their behalf. In the words of Michael McPhearson, the executive director of Veterans for Peace, responding to criticism for the decision to plant fields of crosses in the middle of cities and beaches, 'They say we're not supporting the troops, and they say we shouldn't be doing the vigils. . . . But we feel that especially because we're veterans and we've served, we have the right.'[65] News coverage of events attended and organised by both military families and war vets, tended to accord war veterans a supporting role in relation to Sheehan. Usually they were included along with 'other parents who have lost their children in the war' and 'local activists' as amongst the cast of her fans.[66] This likely worked to their mutual benefit as soldiers did not risk the appearance of mutiny when they were not at the forefront of the charge against their government. Instead their presence and their words served to further authorise the legitimacy of Sheehan's speech, granting her the right to speak on their behalf. The inherited legacy of the citizen-soldier tradition, on which her discursive positionality rested, helped to reach the nation's psyche by invoking patriotic and militaristic values while opposing the war.

THE FIGURE OF THE SOLDIER

This relationship also enabled Sheehan to bolster her motherly position as loyal and fierce protector. She could and did say 'I support the troops' without it sounding like an apology for what she said next. As a spokesperson for the troops and the antiwar movement, Sheehan was able to forcefully argue that if you support the troops (which you must), you should not support this war. There was obviously an inverse gendered logic of protection at work wherein contrary to established ideas about the 'just warrior' protecting the feminised homefront and 'beautiful souls,' the mothers were out doing battle to protect their sons from the military men.[67] This could be described, following Sara Ruddick, as an instance in which the maternal rationality of care came up against a military logic and deemed it perverse. As she explains:

> Mothering begins in birth and promises life; military thinking justifies organized deliberate deaths. A mother preserves the bodies, nurtures the psychic growth, and disciplines the conscience of children she cares for; the military deliberately endangers the same body, mind, and conscience in the name of victory and abstract causes. Mothers protect children who are at risk; the military risks the children mothers protect.[68]

Without denying Ruddick's contention that mothering practice provides women with an alternative standpoint (or standpoints) from which to voice an alternative rationality (or rationalities), my interest is in the inverse: in the subjects who were imagined and enabled by this and similar discursive frames. For instance, as already argued, such an understanding of maternal practice enabled mothers to position themselves (against a cultural legacy of gendered guilt) as loyal nurturers and gentle protectors. In addition to this, it promotes a conceptual dichotomy between the military boys and the military men, casting the soldier in the figure of a child – amongst the innocent.

To the extent that this discursive frame recalls easy oppositions between predators and innocents, it may unwittingly reproduce an image of the 'innocent American soldier' – an image around whom historic tales of American masculine heroism have been built.[69] This is the image often portrayed in the movies and in speeches by presidents. Take, for example, the following quote from President Nixon, in which he contrasts the soldiers fighting in Vietnam with antiwar activists at home:

> Those kids out there (soldiers in Vietnam) . . . I have seen them. They are the greatest . . . [They are unlike the antiwar activists on university campuses] . . . You see these bums, you know, blowing up college campuses today . . . storming around about this issue . . . . Then out there we have kids who are just doing their duty. They stand tall and they are proud . . . . They are going to do fine and we have to stand in back of them.[70]

President Reagan, when honouring the memory of the 50,000 soldiers, who died in Vietnam, offered a similar portrayal: 'They put their lives in danger to help people in a land far away . . . . All were patriots . . . . They were both our children and our heroes. We will never forget their devotion and their sacrifice.'[71] Noteworthy in both quotes is the combination of virtue and innocence that underpins the figure of the soldier in American cultural memory. The military itself is often depicted as a way station or rite of passage in the transition from boyhood to manhood – where one enters a boy, wide-eyed and full of idealism, and comes out a man, with an appreciation of some of the grim realities of this world, including the recognition of his own mortality.[72] The image of the soldier as boy also belies understandings of death in war as the ultimate sacrifice because a soldier who dies at war sacrifices not only his life for his country, but a life not yet lived. Having died at the threshold of manhood, so the stories go, 'he' will never have a family, return to his newlywed wife, or fulfil his childhood dream. His is a life both unfulfilled and full of promise; it is a life that must be grieved.[73]

Boose explains that 'American militarist ideology has been built on pro-tecting [the image of the American soldier as innocent].'[74] At the time of the Gulf War, antiwar sentiment was re-remembered as the cause of the vari-ous harms suffered by Vietnam Vets and was delegitimised on that basis. Says Boose, 'The threat to destroy the signified – the rectitude of US mili-tary action – [was] tantamount to attacking the boy himself.'[75] Or, perhaps, it was tantamount to attacking *the man* or a white masculinity that – as read through the cultural re-emergence of the figure of 'the Vet' (cast in white) in the TV and film productions of the late 1980s – imagined itself as under siege.[76] This time the American antiwar movement somewhat success-fully reversed this logic, not by disregarding or challenging the predominant image of the soldier, but by aligning itself with this figure and suggesting that in the current Iraq War 'being for' the troops means 'being against' the war. As in retrospective accounts of the Vietnam War, the soldier was cast as an innocent victim; what changed was that he was not cast as a victim of the war protester, but of the Bush Administration.

This was a not insignificant shift. As Maureen Dowd, from the *New York Times*, reported, 'The White House used to be able to tamp down criticism by saying it hurt our troops, but more people are asking the White House to explain how it plans to stop our troops from getting hurt.'[77] In addition to the slogan, 'Support our troops. Bring them home now,' Sheehan's question, 'What noble cause?' rallied the mainstream antiwar movement – i.e., what noble cause is worth the sacrifice of the nation's innocents?[78] Sheehan, of course, believed there was none – that the nation had been lied to and that her son and others had been the unwitting victims:

> We all know the reason that we are in Iraq . . . it is for oil. It is so George, Dick, and their evil buddies can rape more profits from our children's flesh and blood. This is not a noble cause, it is the most ignoble cause for a war that has ever been waged.[79]

This protest movement, like many, relied upon that which Michael Blain has called 'victimage rhetoric,' replete with a cast of victims and innocents.[80]

Sheehan and the other mothers who joined her were cast as 'speak-ing truth to power' – as those charged with the responsibility to protect the nation's children and defend human life. On more than one occasion,

Sheehan noted her failure to protect her own son – from the lies of the military recruiter who promised Casey would not see combat and the lies of the Bush administration.[81] But, she nonetheless spoke proudly of Casey's unwavering sense of duty in the face of doubts they both shared about the war in Iraq: 'He said, 'Mom, I have to go to war. It's my duty. My buddies are going.'[82] Casey, of course, represented the quintessential American soldier. Media accounts noted that prior to his military enlistment, he was an Eagle scout (the highest scout ranking) and served as an altar boy for ten years. He died at age twenty-four trying to help others, having volunteered to help retrieve injured soldiers when his own convoy came under attack.[83] Another mom, who joined Sheehan's 'Bring Them Home Now Tour,' described the plight of her own son who was still serving in Iraq at the time, by saying, 'He is so young, so idealistic, so confident, so trusting.'[84] The tragedy of this 'ignoble' war was compounded by the image of those sent to fight – the virtuous and innocent.

THE FIGURE OF THE VIETNAM VET

The intention of this paper is not simply to argue the opposite – to recast soldiers as murderers or misogynists or to offer a more accurate portrayal of soldiers by drawing attention to, amongst other things, the increase in the number of women. My intention, rather, is to illuminate that which compels the socio-cultural demand to pay tribute to this particular image – the image of the soldier as young, naïve, and innocent. In the first sections of this paper, I have outlined some of the narrative resources that preceded and enabled the mainstreaming of the antiwar movement in and through the figure of Cindy Sheehan. I have outlined the way in which Sheehan emerged on the national stage through an alignment with the figure of the soldier – and began mapping this discussion onto the gendered and racialised terrain of American domestic politics to get an image of the new geography of support/dissent that made the war in Iraq intelligible and, for some, unintelligible beyond the battlefield. I have, up until this point, been primarily concerned to illustrate the possibility of dissent in and through the phrase 'I support the troops' as usurped and uttered to profound effect by a grieving mother. In the latter sections of this paper, by contrast, I want to investigate the simultaneous im-possibility of dissent by unpacking more thoroughly the compulsive forms of erasure and forgetting that accompany this utterance – or, more to the point, the unavowable losses that haunt/compel the visible, the incessantly present, the say-able, and the said. To do this is, on the one hand, to investigate those made present in and through the iconic figures who told the nation *who it was* in the disassembling/reassembling post-9/11 period – such that, as I will illustrate, we can link Sheehan not only to the figure of the soldier, but also to the figure of the firefighter who emerged from the wreckage of the Twin Towers. On the other hand, it is to investigate the constitutive absences and unavowable losses that have been formative of *the* and *those* present. Herein, I will suggest, we can trace a culturally instituted, discursive form of melancholia that has shaped the realms of possible and impossible dissent – and has done so precisely *to the extent* that it has governed the possibility of intelligible subject formation.

What is notable about the image of the soldier figure, for instance, is the extent to which it is, in many ways, the opposite of the image of the emasculated and crippled Vietnam Vet – the one who, having become remembered as the victim of antiwar protest, arguably spurred the almost manic need to demonstrate support for the troops in the Gulf War. Although the legacy of betrayal associated with this figure seemingly lives on today, haunting the 'support the troops' discourse of present, he is also effaced in this slogan's every utterance. He has become meaningful precisely through his absence. Not only has his face been eclipsed by that of a fresh, clean-cut looking recruit, but judging by the number of homeless and drug-addicted Vietnam Vets, he is not the referent object of the nation's desire to support and protect.

Although retrospectively cast as the victim of a castrating woman, flag-burning scum, or another form of evil 'other,' the varied responses accorded to this figure and the struggles to write and rewrite his place in the nation's history suggest a much broader collective ambivalence. Unlike the image of the soldier, the injured figure of the Vietnam Vet represents failed boyhood promise. As in the movies, *Born on the Fourth of July* and *Forest Gump*, he emerged from the military as a broken man – physically, psychologically, and morally. He represents a figure of betrayal – one who was both betrayed by a nation that could not easily reconcile his image with tales of masculine heroism and one who betrayed the nation, as he will forever be associated with the nation's military and moral failure – its impotence and lack. In popular portrayals the figure of the disabled Vietnam Vet came to stand in for the failures of the nation itself – as a visible blight on what Robert McMahon refers to as 'the national narrative of progress, achievement, and righteousness.'[85] The nation, by extension, was sometimes portrayed in popular culture and presidential speeches in terms of an emasculated man – imagined as traumatised or as suffering a debilitating disease.

Typically, the nation was portrayed this way by the same people who wanted to rejuvenate it and restore it to health by re-invigorating its sense of purpose (which was often envisioned in terms of its military prowess abroad). In the words of President Ford (1975), for example, 'The time has come to look forward to an agenda for the future, to unity, to binding up the nation's wounds and restoring it to health and optimistic self-confidence.'[86] Similarly, in a 1997 *Wall Street Journal* op-ed column, Ernest Lefever took aim at antiwar critics whom he accused 'of continuing to harbor negative views not just toward the Vietnam intervention but toward America as well.'[87] In his words:

> The cynical view of our involvement in Vietnam became part of a larger culture of shame, guilt, and self-flagellation that erupted in flag-burning and other attacks on traditional institutions. It also helped spawn the 'Vietnam Syndrome' that all but paralyzed America from using military force abroad.[88]

Views such as these tended to obfuscate what McMahon refers to as 'widespread feelings of anguish, revulsion, and opprobrium toward the whole Vietnam experience.'[89]

The ontological insecurity (the crises of identity, meaning, and values) – indeed, the profound sense of disillusionment and loss – that the episode in Vietnam reportedly generated was trivialised and delegitimised by the health metaphors that likened the experience to an unfortunate, but temporary, disease that the nation could and must shake as one would a cold.[90] Rather than engaging in a moral reckoning or coming to terms with their accountability for the events of Vietnam, Americans were urged to close the chapter on Vietnam – to cast off their anguish and self-doubt, lest they erode the nation's overall sense of purpose.[91] In return, it was promised that the nation's virtue, innocence, and virility – even ontological security – could be restored.

But, buying into this promise meant casting off parts of the self as a 'defiling otherness'[92] – attributing one's own anger and misgivings to an 'angry,' 'troop-hating,' and 'flag-burning' antiwar protester, re-imagined as the cause of the affliction that the nation was suffering. This transference and the complex discursive maneuvering that accompanied it enabled the Vietnam Vet and the nation itself to occupy the ground of victim again. Boose captures this process well:

> . . . as the nation reacted against the national guilt that it had tentatively begun to confront in the 1970s, the only image that America "saw" by the late 1980s was the convincing picture of itself as the proverbially innocent U.S. soldier, returning from war and being victimized by the insults of 'spitting and jeering throngs.'[93]

This was in spite of the fact that polls taken from 1969 onwards suggest that the majority of Americans have continued to harbour negative views of the Vietnam War. A 1982 poll, for instance, found that 72% of Americans considered the war 'fundamentally wrong and immoral' and more recent polls indicate that little has changed since that time.[94] Furthermore, the legacy of suicide, homelessness, and addiction that has afflicted Vietnam Vets suggests that society has never fully come to terms with the psychologically and/or physically crippled figure of the Vietnam Vet or cast *him* within its protective net. This figure, rather, is a living ghost that haunts the pledge of *never again* and the compulsory demand to *this time* support our troops – even as he is effaced in the pledge's every utterance.

THE FIGURE OF THE FIREFIGHTER

If in the 1980s, the impulse was to repeat and re-do,[95] to effectively rewrite the legacy of Vietnam and to maniacally refuse the task of grieving or the process of coming to terms with its shame and guilt, in the immediate aftermath of 9/11, the collective impulse was simply to forget and to leap-frog over that ripple in the nation's history. The terrorist attacks on the Pentagon and World Trade Center exposed American vulnerability and for a moment, at least, enabled the nation to not only occupy the site of victim, but to recapture its mythicised, but badly tarnished, innocence. The repeated refrain, *Why do they hate us so much?*, contained a tacit recognition of the reason for the attacks coupled with a simultaneous protest of innocence. 'Symbolically,' says Howard Stein, 'America became the site of the evil for the enemy, and the place where Americans struggled for a renewed sense of goodness.'[96]

Many cultural commentators have noted that a nostalgic longing for unity, purposefulness, and innocence was in evidence prior to the 9/11 attacks – in the country's renewed interest in World War II evidenced in popular movies – such as *Saving Private Ryan* (1998); *Pearl Harbor* (2001); and the HBO miniseries, *Band of Brothers* (2001) – and in the popularity of Tom Brokaw's book, *The Greatest Generation* (1998).[97] Hence, despite the novelty of the terrorist attacks, certain mythical frameworks already existed within which the events of that day could be interpreted and assigned cultural meaning and significance. As Stein explains, 'When the attacks in New York City and Washington, D.C. happened, we . . . already 'knew' much of what they signified . . . . The popular 2001 movie *Pearl Harbor* was still active in the cultural memory.'[98] While I think there were many competing narratives jostling to capture the imagination of the nation, certainly the legacy of World War II was prominent among them and, as Stein suggests, unconsciously replayed in the cultural imaginary:

> When the heroic firefighters planted the American flag in the rubble of the World Trade Center, one immediately 'knew' – recognized via projection – that a reenactment of the Asian Theatre in WWII was taking place. The firefighters were immediately likened to the marines who raised the American flag on Mt. Suribachi, Iwo Jima, on February 23, 1945. Past was merged with present.[99]

Arguably, the exposure of American vulnerability in the face of surprise attacks on its own soil enabled a symbolic return to an imagined simpler time, pre-Vietnam, when the nation was on the cusp on manhood – wide-eyed, virtuous, and innocent – and when the boys who returned from war, returned having successfully transitioned to manhood – as honourable, stoic, and emotionally grounded men.[100]

Of course, the ghosts of Vietnam still lingered here. Recollections of World War II as 'the good war' – 'as one of the nation's most glorious and righteous moments' – stood in uneasy contrast to the 'antithetical status' accorded to the war in Vietnam.[101] In the course of the nation's symbolic return, the experience of Vietnam was yet again vigorously disavowed – this time relegated to the realm of 'never, never' where what never was, was never lost such that the mythic nation could go on.[102] This is what Butler, quoting Hegel, would refer to as 'the loss of the loss.'[103] This was the melancholic structure within which Sheehan emerged – where disavowed loss came to govern what could be, even what had to be, said and the forms of loss that could be recognised and mourned. Stein writes that '[a] large part of the work of culture "itself" is an embodiment and enactment of the inability to mourn.'[104] Ontological certainty was (re)founded upon a mythicised past and certain attachments and ideals that seemingly had to persist as they informed who Americans imagined themselves to be, their place in the world, and their constitutive emotions and desires.

Thus, the firefighters, the big burly men who emerged from the wreckage of the World Trade Center covered with ashes and soot, were imagined in terms of World War II's *greatest generation*: 'brawny, heroic, manly men'.[105] In the words of conservative social critic Camille Paglia, 'I can't help but noticing how robustly, dreamily masculine the faces of the firefighters are. These are working-class men, stoical, patriotic. They're not on Prozac or questioning their gender.'[106] Explaining this 'longing for manliness,' David Granger, the editor-in-chief of *Esquire* magazine, explained, 'Now there's a sense of selflessness being attributed to rugged men . . . . People want to regain what we had in World War II. They want to believe in big, strapping American boys.'[107] According to a *New York Times* social commentary, unlike the terrorist masculinity that was 'breeding in some cultures,' what occurred in the United States was the emergence of a 'tough-tender hero.'[108] This hero figure was imagined to embody the goodness and the promise of America – an America wherein boys could be boys and men could be men and the re-affirmation of this was seen, in and of itself, to signify renewal, wholesomeness, and security.[109] Noting the cultural homage that has been paid not only to fire-fighters, but also to police and emergency workers, to the men who tackled the hijackers on United Flight 93, and to the welders and excavators who dug through the wreckage of the WTC, George Packer of the *New York Times* writes of the 'redemption' of working-class men in the post-9/11 period: 'Those workers who were fast becoming relics turn out to have preserved the only qualities that matter when crunch time comes.'[110] As he and others have described, these are the qualities of stoicism, brotherly sacrifice, and good, honest work.[111]

Unlike the figure of the crippled Vietnam Vet and unlike the much more boyish figure of the soldier, these men fully and successfully transitioned to an idealised (and dreamy) manhood. They were imagined as neither objects of pity mixed with revulsion, nor as precarious and in need of our protection. They were the figures Americans were expected to pay homage to and the figures that women fantasised about. Never mind the details that did not coincide with the fantasy. In an article written for the *New York Times*, N. R. Kleinfield wrote that an internal inquiry conducted by the New York Fire Department documented the fear, panic and confusion that firefighters expressed and a rescue mission that was, in many respects, in disarray.[112] But, while acknowledging that the collected transcripts and interviews present a more complicated picture of what transpired than early media accounts – a picture of both human strength and fallibility – he concluded that 'none of this dilutes the valor of *the men* who rushed into the towers . . . [in fact, it] only deepens an appreciation of their courage.' Kleinfield pointed to the limitations of the original version of events, but set his own limits on the possibilities for re-interpretation – specifically leaving a mythical image of the heroic, male firefighter intact. Never mind, as Carol Gilligan points out, that some of the heroic firefighters and police officers who rose to the occasion happened to be women.[113] Never mind that the greatest generation (whose legacy is recalled in the image of the firefighter) was not always imagined so – that those who protested against the Vietnam War often extended their criticism to encompass the values and way of life associated with the World War II era and the postwar years. In fact, never mind the impossibility of all these figures – the stoic firefighter with the protestant work ethic, the spat-upon Vet, or the wide-eyed, innocent soldier. The point is that when President Bush declared that 'another great generation' had been summoned to act, a line of continuity was invoked with a mythicised past and a nation was (re)founded in spite of the events and questions (*Why do they hate us so much?*) that threatened to tear it apart.[114]

FROM THE POSSIBILITY TO THE IMPOSSIBILITY OF DISSENT (AND BACK AGAIN?): CONCLUDING REMARKS

Against this background, Sheehan's question, 'What noble cause?', was particularly prescient. As one person, quoted in the *New York Times*, put it when asked to explain why he did not want to see the movie *Flags* even though he loved the World War II–based book:

> It was possible to look back at World War II with nostalgia, and think that those were great men doing great things that Americans would never have to do again. You'd think, well, people were shot to bits, but that was then . . . . Now . . . you know it's happening in Iraq . . . and for a lot less noble cause.[115]

Sheehan highlighted the simple fact that this was not World War II all over again and yet, as with Kleinfield, she demanded that certain elements of the original narrative be allowed to persist. In fact, the rallying cry to 'support the troops,' which bridged the pro-war and anti-war divides, invoked – repeatedly – the possibility of harm that an imaginary 'we' can do to 'them' by refusing this call. Motherhood, in this particular instance, afforded Sheehan and the antiwar protesters, who variously aligned themselves with her, a platform in which misgivings about the war could be articulated and comparisons with World War II could be negated while attachments and identities forged in and through alignments with the soldier and the firefighter could persist. Indeed, the latter was a condition of possibility for legitimate antiwar protest – as well as that which marked the realm of disqualified speech.

This is not to suggest that Sheehan's critiques never fell out of bounds; they did, threatening to destabilise the very boundaries that enabled her to be a spokesperson for the antiwar movement. But, she was successful to the extent that her identity as a grieving mother and loyal nurturer, first and foremost, came to matter more than *the rest*. It was the condition that qualified her speech, delimiting what was heard. To repeat David Letterman, "How can [anyone] possibly take exception with the motivation and the position of someone like Cindy Sheehan?"[116] The audience could applaud Letterman and rally behind Sheehan without engaging in a serious conversation about her designation of the label 'freedom fighter' to Iraqi combatants – a conversation that could potentially destabilise idealised images of our soldiers and firefighters, but that was actually intended to disparage her by challenging her primary allegiance to them. By indirectly appealing to her identity as a grieving mother, Letterman reaffirmed precisely this, granting her speech legitimacy even while inadvertently disqualifying what she said. This was not World War II all over again and yet the image of the idealised national community associated with World War II was made to persist. Within Sheehan's plea to the nation (to support the troops by urging that they be brought home), Americans were asked to identify with a mother's need to protect a revered, yet still vulnerable, masculinity – which became inextricably bound with American goodness and innocence. By supporting the troops, even if not the war, a line of continuity with a mythic past was not severed and a deeper questioning of American cultural values, consumerism, militarisation, and gender norms was evaded. The troops and the firefighters came to stand in for an ideal or attachment that seemingly could not be mourned or protested against as the attachment structured hope and desire and the identities that were generated through them. By saving them, the nation saved itself.[117]

This tells us something about the psychic life of power – the intimacies of Empire and even its architecture. The political – and, indeed, even the geopolitical configurations of our world – are personal and we do not exist outside of its structures. The problem is not just that we have 'all these theories . . . yet the bodies keep piling up' (to borrow Marysia Zalewski's phrase),[118] but that the bodies disappear from view – even in the rallying cries of those protesting the war. The aim of this paper has been to show that while the mother of a dead soldier usurped the phrase 'I support the troops' to galvanise an antiwar movement, this compulsory utterance was a product of a socio-cultural melancholia that worked to structure *the visible, the sayable, the said* and *the heard* – all of which were underpinned by compulsive erasures. The latter section of this paper, in particular, has endeavoured to *show* those who have been constitutive in their absence. There were those such as the women firefighters, for instance, who were rendered invisible to enable the re-emergence of a heroic masculinity – as read through the bodies of men. There were *the assorted discontents* (amongst whom we could include veterans, mothers, fathers, soldiers, rock stars, and other conscientious citizen-subjects opposed to the war) who were silenced or who *spoke out*, but in throttled ways, in order to be heard at all. Not least importantly, there were those whose lives and communities were destroyed – and those who were maimed or killed and then erased – in order to sustain the nation's (re)founding myths. These were the bodies that were spoken out of discursive existence or spoken into it only to be aggressively effaced. Collectively, these *are* the bodies that have been vigorously disavowed and yet relentlessly persist.

These disavowals continue to haunt the American cultural landscape. My aim throughout has been to cast it too as a space of war – one that is fundamentally constituted in violence, but marked by the simultaneous erasure of this violence (or, at least, the denial of this violence as stemming from within). The home-front is just that – one of the fronts in the battle for Iraq (and one that is never totally outside it) as those who wage war know all too well (as they admonish a less-than-enthusiastic public for diminishing soldiers' morale and, well, for making the practice of warfare just plain uncomfortable and difficult). The challenge in writing this paper has been to make the war beyond the battlefield visible – by pointing not only to the bodies made present within the demand to support the troops, but also by pointing to the bodies who are constitutive of this socio-cultural imperative precisely in their absence (those, in other words, who are compulsively *made absent*). Paradoxically, in so doing, I hope to destabilise the field of representation that makes the soldiers, the firefighters, and even Cindy Sheehan possible (at least as culturally recognisable and understood) – by pointing to their fundamental impossibility.

Butler explains that for Emmanuel Levinas 'the human is indirectly affirmed in the very disjunction that makes representation impossible.'[119] Arguably, the iconic image (Figure 1) of the Vietnamese girl who was caught on film running down the street naked and 'screaming in pain and terror from the napalm burns on her back and arms,' jarred the nation's collective

FIGURE 1 The Girl in the Picture.

psyche to the extent that her image constituted a rupture in the hegemonic field of representation.[120]

Robert Hariman and John Louis Lucaites have argued that for a moment at least, what the nation *saw* in this photograph was not just a Vietnamese peasant girl (whatever that had come to mean), but a fundamental humanity constituted in vulnerability – casting the American viewer in an 'elemental moral situation': 'Like the parable of the Good Samaritan, which featured a naked man discovered along a road, the girl's naked vulnerability is a call to obligation, and, as in the parable, one that has occurred unexpectedly.'[121] The girl's image *worked* not by producing a relation of sympathy (which may leave the hegemonic field of representation intact), but to the extent it produced a moment of radical de-centring as the American viewer became reconstituted through its relation to her. Arguably herein lay the photograph's pivotal impact – its ability to re-direct feelings of 'towardness' and 'awayness' to/from the Vietnam War.[122] This was also a moment of expansive possibility – a calling to accountability that, if heeded, would demand a forging of new identities, meanings, and narratives as well as the acceptance of a certain degree of ontological insecurity, humility, and the painful recognition of America's lost innocence and heroic sense of destiny.

The iconic image of the 'girl in the picture' collided with official representations of the Vietnam War and the noble cause that American soldiers were involved in; representation broke allowing her to be seen and heard. As of yet, there is no iconic image from the Iraq War that has jolted the collective imaginary in quite the same way. Even the pictures from Abu Ghraib have seemingly failed to really de-centre hegemonic representations – especially in terms of the identities involved which have largely remained stable.

In *Precarious Life: The Powers of Mourning and Violence* Butler asks, 'How does the prohibition on grieving emerge as a circumscription of representability, so that our national melancholia becomes tightly fitted within the frame of what can be said, what can be shown [and, I would add, what can be seen and heard]?'[123] For the images of destruction and death and the statistical estimates of lives lost that have come to us from Iraq are of little import if the lives lost are merely that – an unfortunate statistic or fact amongst the many other unfortunate facts of this world.

The compulsive demand to support the troops even if not the war, reveals a melancholic structure of desire in which an attachment to lost ideals (an innocence which cannot be restored) and a lost sense of purpose or place in the world (based on an imaginary unity from a time past) superseded any desire to reckon with shameful elements of the past and to significantly alter course or re-become. The nation, as a result, was compelled to repeat and re-do. What could be seen and heard were circumscribed by that which seemingly *had to be* if the mythic nation and the identities that sustained it were to persist. Herein lay the im/possibility of dissent, but also the possibility of re-presentation – a possibility which resides in the impossibility of representation itself. When Dexter Filkins of the *New York Times* described the Iraqi reaction to the successful American occupation of Baghdad in the first few days of the war, he noted that in addition to the handful of small groups of demonstrators celebrating the American victory, there were a number of Iraqis who came out to watch the events unfold, but who stood back and watched silently, apparently 'unwilling to participate.'[124] Whether this be from 'lingering fear or pain [of the Iraqi regime],' he said, 'was not clear.' He characterised this in terms of a muted euphoria. In his exact words, 'There was a muted quality to the euphoria today'[125] – the impossibility of which strikes me as profanely obscene.

However, within the limits of this representation, within the limits of available narrative frames, and within the disjuncture between this scene and its characterisation, possibility exists in the simple fact that what *appears* is not an image of an Iraqi (whatever that might mean), but the figure's radical impossibility. As Butler explains, following Levinas, 'For representation to convey the human . . . representation must not only fail, but it must *show* its failure.'[126] Within this disjunction then a new and potentially expansive space emerges to pursue alternative relations and ways of being, but with it emerges a new and potentially painful call to accountability. Heeding this call would demand not only coming to terms with the impossibility of the Iraqi, but also all the other figures that have been imagined alongside him/her to perpetuate the nation's (re)founding myths. It would demand a letting go of constitutive desires and truisms forged according to regressive fantasies and a coming to terms with uncertainty, vulnerability, and the fact of one's own alterity.

ACKNOWLEDGEMENTS

I would like to thank David Mutimer, the anonymous reviewers, and the editors of this special edition (David Grondin and Simon Dalby) for the thoughtful feedback and insights offered along the way.

NOTES

1. I took the phrase 'loss of the loss' from Judith Butler, who is quoting Hegel, in J. Butler, *The Psychic Life of Power* (Stanford, CA: Stanford University Press 1997) p. 24. Also see her interview in V. Bell, 'On Speech, Race and Melancholia: An Interview with Judith Butler', *Theory, Culture, and Society* 16/2 (1999) p. 171.

2. M. Foucault, 'Politics and the Study of Discourse', in G. Burchell, C. Gordon, and P. Miller (eds.), *The Foucault Effect: Studies in Governmentality* (Chicago: Chicago University Press) p. 72.

3. See J. Butler, *Excitable Speech: A Politics of the Performative* (Routledge: New York 1997) pp. 5–34; and Butler, *Psychic Life of Power* (note 1) pp. 20–21.

4. See Butler, *Psychic Life of Power* (note 1)

5. Ibid., pp. 20–21.

6. J. Butler, *Precarious Life: The Powers of Mourning and Violence* (New York: Verso 2004) pp. 22–24.

7. See, for example, Butler, *Psychic Life of Power* (note 1) pp. 185 and 196. Also see W. Brown, 'Resisting Left Melancholy', *Boundary 2* 26/3 (1999) pp. 19–97.

8. See, for example, Butler, *Precarious Life* (note 6) p. xiv.

9. Butler, *Psychic Life of Power* (note 1) p. 170. Also see pp. 185–186

10. See, for example, R. Salladay, 'Anti-War Patriots Find They Need to Reclaim Words, Symbols, Even U.S. Flag from Conservatives', *San Francisco Chronicle* (7 April 2003).

11. Quoted in ibid.

12. See the following website: <Bringthemhomenow.com>, accessed 24 April 2007.

13. See Salladay (note 10).

14. See, for example, E. Bumiller, 'In the Struggle over the Iraq War, Women are on the Front Line', *New York Times* (29 Aug. 2005).

15. Ibid.

16. CodePink: Women for Peace, Newsletter, 10 Aug. 2005.

17. C. Sheehan, *Peace Mom: A Mother's Journey through Heartache to Activism* (New York: Atria Books 2006) pp. x–xi.

18. Ibid.

19. W. Brown, 'Freedom's Silences', in R. C. Post (ed.), *Censorship and Silencing: Practices of Cultural Regulation* (Los Angeles: The Getty Research Institute for the History of Art and the Humanities 1998), quotes on pp. 320 and 315. See entire article for more details, pp. 313–327.

20. The expression 'freedom's intransigence' borrows from M. Foucault's expression, 'the intransigence of freedom.' For more, see M. Foucault, 'The Subject and Power', in P. Rabinow and N. Rose (eds.), *The Essential Foucault* (New York: The New Press 2003) p. 139.

21. Butler, *Excitable Speech* (note 3) p. 51.

22. Ibid., p. 16.

23. R. H. King quoted in T. Lovell, 'Resisting with Authority: Historical Specificity, Agency, and the Performative Self', available at <http://www.socialsciences.manchester.ac.uk/sociology/Seminar/documents/terrybourdieupaper.doc>, p. 9, accessed 23 Aug. 2006. Longer version of paper also available in *Theory, Culture, and Society* (2003).

24. 'One Mother in Crawford', *New York Times*, 9 Aug. 2005. Also see the following: H. Hugg, 'The Conscience of the King', *Guernica Magazine* (Sep. 2005), available at <http://www.guernicamag.com/features/90/the_conscience_of_the_king/>, accessed 19 July 2006; S. G. Stolberg, 'A Year after March against Iraq War, Another Try', *New York Times*, 7 Aug. 2006; A. Ripley, 'A Mother and the President', *Time Magazine*, 15 Aug. 2005; and R. Stevenson, 'Of the Many Deaths in Iraq, One Mother's Loss Becomes a Problem for the President', *New York Times*, 8 Aug. 2005.

25. Ripley (note 24).

26. E. Bumiller, 'Turning Out to Support a Mother's Protest', *New York Times*, 18 Aug. 2005.

27. Ibid.

28. Lovell (note 23).

29. Ibid., p. 7.

30. Ibid., p. 5.

31. Ibid., p. 8.

32. CNN, *CNN Transcript*, 15 Aug. 2005, available at <http://transcripts.cnn.com/TRANSCRIPTS/0508/15/acd.01.html>, accessed 19 July 2006.

33. Ibid.

34. Hugg (note 24).

35. CodePink (note 16).

36. H. Anderson, 'As Debate Rages, Camden Opts for "Yellow Ribbon Panel"', *Camden Village Soup Times*, 3 June 2003; 'Peninsula Dons Yellow Ribbons: Mother Post Yellow Ribbons on Streetlights', *NBC11.com* (serving San Jose, Oakland, and San Francisco), 31 March 2003, retrieved 9 June 2004 from NBC11.com; S. Stevens, 'Moms Want Ribbons on Area Trees', *Daily Herald*, 20 March 2003; 'Town Protests Mayor's Ban on Yellow Ribbons: Marchers Hit Streets to Tack Emblems on Cars, Buildings, Trees', *WorldNetDaily.com*, 5 April 2003, retrieved 10 June 2004 from worldnetdaily.com.

37. Anderson (note 36).

38. 'Town Protests Mayor's Ban' (note 36).

39. Ibid.

40. Hugg (note 24).

41. See, for example, Ripley (note 24).

42. For more see the following: J. Butler in Bell (note 1); P. Hill Collins, 'It's All in the Family: Intersections of Gender, Race, and Nation', *Hypatia* 13/3 (Summer 1998) pp. 62–81; N. Yuval-Davis, Women and the Biological Reproduction of "The Nation"', *Women's Studies International Forum* 19/1–2 (1996) pp. 17–24.

43. Hill Collins (note 42) p. 77.

44. E. Bumiller, 'Bush and the Protester: Tale of Two Summer Camps', *New York Times*, 22 Aug. 2005; and E. Peretz, 'High Noon in Crawford', *Vanity Fair*, Nov. 2005.

45. Stevenson (note 24).

46. Ripley (note 24).

47. A. K. Brown, 'Grieving Mother's War Protest Draws Notice', *Associated Press*, 11 Aug. 2005.

48. D. Stout, 'Bush tells Veterans that Iraq Policy Will Make U.S. Safer', *New York Times*, 22 Aug. 2005.

49. This ban was originally put in place during the first Gulf War by George Bush Sr. – ostensibly for the purpose of protecting the privacy of grieving families. President Obama has rescinded the outright ban, leaving it to the discretion of individual families to determine whether or not they would object to the presence of media cameras.

50. Brown, 'Freedom's Silences' (note 19) p. 315.

51. Ripley (note 24).

52. Ibid.; R. Moore, 'Conservative Media Smearing Cindy Sheehan', *LewRockwell.com*, 12 Aug. 2005, available at <http://www.lewrockwell.com/orig6/moore-r2.html>, accessed 7 June 2006; E. Bumiller, 'In the Struggle over the Iraq War, Women are on the Front Line', *New York Times*, 29 Aug. 2005.

53. 'O'Reilly on Cindy Sheehan', *Media Matters for America*, 6 Jan. 2006, available at <http://mediamatters.org/items/200601060009>, accessed 3 Jan. 2007.

54. E. Morris, 'Conservative Compassion', *New York Times*, 17 Aug. 2005.

55. Ripley (note 24); Brown, 'Grieving Mother's War Protest' (note 47); Town Forum Press, *Letters to Cindy Sheehan: Messages to the Left on America's Noble Cause in Iraq* (Fairfax: Town Forum Press 2006); A. Goodnough, 'In War Debate, Parents of the Fallen Are United Only in Grief', *New York Times*, 28 Aug. 2005.

56. P. DiQuinzio, 'Love and Reason in the Public Sphere: Maternalist Civic Engagement and the Dilemma of Difference', in S. M. Meagher and P. DiQuinzio (eds.), *Women and Children First: Feminism, Rhetoric, and Public Policy* (Albany: State University of New York Press 2005) p. 240.

57. Idea of 'maternal thinking' taken from S. Ruddick, *Maternal Thinking: Toward a Politics of Peace* (New York: Ballantine Books 1989).

58. C. Sheehan, 'Go Home and Take Care of Your Kids', *truthout.org*, 22 Aug. 2005, available at <http://www.truthout.org/cindy.shtml>, accessed 22 June 2006.

59. J. Kovacs, 'Cindy: Terrorists "Freedom Fighters"', *WorldNetDaily.com*, 23 Aug. 2005, available at <http://www.worldnetdaily.com/news/article.asp?ARTICLE_ID=45938>, accessed 3 Jan. 2007.

60. 'O'Reilly on Cindy Sheehan' (note 53).

61. Sheehan also took up the cause of Iraqi mothers and children, but to little notice in the mainstream media.

62. L. Boose, 'Techno-Muscularity and "the Boy Eternal": From the Quagmire to the Gulf', in M. Cooke and A. Woollacott (eds.), *Gendering War Talk* (New Jersey: Princeton University Press 1993) p. 80.

63. Goodnough (note 55); Peretz (note 44).

64. N. Banerjee, 'Iraq Veterans turn War Critics', *New York Times*, 23 Jan. 2005.

65. J. McKinley 'Homemade Memorial is Stirring Passions', *New York Times*, 3 Dec. 2006.

66. M. Santora, 'Mother who Lost Son Continues Fight against War', *New York Times*, 19 Sept. 2005; M. Levenson, 'Mother of Soldier Killed in Iraq Touring Nation', *The Boston Globe*, 18 Sept. 2005.

67. J. B. Elshtain, *Women and War* (New York: Basic Books, Inc. 1987) pp. 3–13.

68. S. Ruddick, 'The Rationality of Care', in J. B. Elshtain and S. Tobias (eds.), *Women, Militarism, and War* (Savage, MD: Rowman and Littlefield 1990) p. 240.

69. Boose (note 62) p. 78.

70. T. D. Beamish, H. Molotch, and R. Flacks, 'Who Supports the Troops?: Vietnam, The Gulf War and the Making of Collective Memory', *Social Problems* 42/3 (1995) p. 352.

71. R. J. McMahon, 'Contested Memory: The Vietnam War and American Society, 1975–2001', *Diplomatic History* 26/2 (2002) p. 168.

72. Boose (note 62) pp. 90, 94; C. V. Scott, 'Rescue in the Age of Empire: Children, Masculinity, and the War on Terror', in K. Hunt and K. Rygiel (eds.), *Engendering the War on Terror: War Stories and Camouflaged Politics* (Burlington: Ashgate 2006) p. 105; S. Whitworth, 'Militarized Masculinities and the Politics of Peacekeeping: The Canadian Case', in K. Booth (ed.), *Critical Security Studies and World Politics* (Boulder, CO: Lynne Rienner 2005) pp. 102–103.

73. The notion of a distinction between 'grievable' and 'ungrievable' lives is taken from Butler, *Precarious Life* (note 6).

74. Boose (note 62) pp. 78–79.

75. Ibid., p. 79.

76. For more on this, see the following: Boose (note 62); S. Jeffords, *The Remasculinization of America: Gender and the Vietnam War* (Indianapolis: Indiana University Press 1989) pp. 117–118 and 120–121; and T. Managhan, '(M)others, Biopolitics and the Gulf War', in S. Meagher and P. DiQuinzio (eds.), *Women and Children First: Feminism, Rhetoric and Public Policy* (New York: State University of New York Press 2005) pp. 205–225.

77. M. Dowd, 'Why No Tea and Sympathy?', *New York Times*, 10 Aug. 2005.

78. Goodnough (note 55); S. Bannerman, 'The Spiritual Crisis in Our Lives Generated by the War in Iraq', *Gold Star Families for Peace*, 19 June 2006, retrieved 22 June 2006 from gsfp.org.

79. C. Sheehan, 'What Noble Cause?', *truthout.org*, 17 Sept. 2005, available at <http://www.truthout.org/cindy.shtml>, accessed 22 June 2006.

80. M. Blain, 'Power, War, and Melodrama in the Discourses of Political Movements', *Theory and Society* 23/6 (1994) pp. 804–837.

81. See Ripley (note 24); Peretz (note 44); Brown, 'Grieving Mother's War Protest' (note 47); C. Sheehan, 'Volunteers for Endless War', *truthout.org*, 22 Aug. 2006, available at <http://www.truthout.org/cindy.shtml>, accessed 21 Jan. 2008.

82. Casey Sheehan quoted in Brown, 'Grieving Mother's War Protest' (note 47).

83. For media portrayals of Casey Sheehan see Peretz (note 44); Ripley (note 24); O. Dorell, 'Soldier's Mother Keeps Protest Vigil', *USA Today*, 6 Aug. 2005.

84. P. Logue, 'Ohio Caravan with "Bring Them Home Now Tour"', *truthout.org*, 12 Sept. 2005, available at <http://www.truthout.org/cindy.shtm>, accessed 21 May 2006.

85. McMahon (note 71) p. 165.

86. Ibid., p. 164.

87. E. Lefevre, quoted in ibid., p. 174.

88. Ibid.

89. McMahon (note 71) p. 175.

90. Ibid., p. 165; S. Niva, 'Tough and Tender: New World Order Masculinity and the Gulf War', in M. Zalewski and J. Parpart (eds.), *The 'Man' Question in International Relations* (Boulder and San Francisco: Westview Press 1998) p. 109.

91. See McMahon (note 71).

92. D. Campbell, explaining Butler's argument, in *Writing Security: United States Foreign Policy and the Politics of Identity*, rev. ed. (Minneapolis: University of Minnesota Press 1998) p. 8.

93. Boose (note 62) p. 84.

94. McMahon (note 71) p. 175.

95. Boose (note 62) p. 75.

96. H. Stein, *Beneath the Crust of Culture: Psychoanalytic Anthropology and the Cultural Unconscious in American Life* (New York: Rodopi 2004) p. 12.

97. See: ibid., p. 6; McMahon (note 71) pp. 177–179; D. Halbfinger, 'Burying Private Ryan', *New York Times*, 29 Oct. 2006; R. Marin, 'Raising a Flag for Generation WWII', *New York Times*, 22 April 2001.

98. Stein (note 96) p. 5.

99. Ibid., p. 6.

100. Concept of 'symbolic return' borrowed from Stein (note 96) p. 16.

101. McMahon (note 71) pp. 177–178.

102. Concept of 'never, never' in terms of disavowed loss taken from Butler, *Psychic Life of Power* (note 1).

103. Ibid., p. 24.

104. Stein (note 96) p. 110.

105. P. L. Brown, 'Heavy Lifting Required: The Return of Manly Men', *New York Times*, 28 Oct. 2001.

106. Quoted in ibid.

107. Ibid.

108. Ibid.

109. J. E. Barnes, 'Heroic Rescue Fighters May Top Holiday Wish Lists', *New York Times*, 26 Sep. 2001.

110. G. Packer, 'Into their Labour', *New York Times*, 1 Sep. 2002.

111. Ibid.; Brown, 'Heavy Lifting Required' (note 105); Marin (note 97).

112. N. R. Kleinfield, 'The Real Story: Heroes, but Human; Firefighters' Accounts Give Sept. 11th the Complex Nuances of History', *New York Times*, 3 Feb. 2002.

113. Noted in Brown, 'Heavy Lifting Required' (note 105).

114. Quoted in Halbfinger (note 97). The particular formulation of this statement is indebted to Whitworth (note 72) p. 89. It should also be noted that the attitude described here echoes the long-shattered and mythicised republican ideal of soldiering explained in Florian Olsen's contribution to this special issue, which nevertheless must remain constantly reactivated to make the sacrifice bearable in the nation's eyes.

115. Halbfinger (note 97).

116. 'O'Reilly on Cindy Sheehan' (note 53).

117. I borrow much of this particular formulation from S. Razack, *Dark Threats and White Knights: The Somalia Affair, Peacekeeping, and the New Imperialism* (Toronto: University of Toronto Press 2004) p. 25.

118. M. Zalewski, 'All these Theories Yet the Bodies Keep Piling Up', in S. Smith, K. Booth, and M. Zalewski (eds.), *International Theory: Positivism and Beyond* (Cambridge: Cambridge University Press 1996) pp. 340–353.

119. Butler, *Precarious Life* (note 6) p. 144.

120. Quote from R. Hariman and J. L. Lucaites, 'Public Identity and Collective Memory in U.S. Iconic Photography: The Image of 'Accidental Napalm', *Cultural Studies in Media and Communication* 20/1 (2003) p. 53. Also see Butler, *Precarious Life* (note 6) p. 150. The iconic photograph featured here was taken by Associated Press photographer Nick Ut on 8 June 1972. The image and further details were retrieved on 22 Oct. 2010 from <http://en.wikipedia.org/wiki/Nick_Ut>.

121. Hariman and Lucaites (note 120) p. 42.

122. Expression of 'towardness' and 'awayness' taken from Sarah Ahmed, *Cultural Politics of Emotion* (New York: Routledge) pp. 7–8.

123. Butler, *Precarious Life* (note 6) p. 148.

124. D. Filkins, 'People Rise as Icons of Nation Fall Down', *New York Times*, 10 April 2003.

125. Ibid.

126. Butler, *Precarious Life* (note 6) p. 144 (italics in the original).

Index

Page numbers in **Bold** represent figures.

Related titles from Routledge

A Micro-Sociology of Violence
Deciphering Patterns and Dynamics of Collective Violence

Edited by Jutta Bakonyi and Berit Blieserman de Guevara

This book aims at a deeper understanding of social processes, dynamics and institutions shaping collective violence. It argues that violence is a social practice that adheres to social logics and, in its collective form, appears as recurrent patterns. In search of characteristics, mechanisms and logics of violence, contributions deliver ethnographic descriptions of different forms of collective violence and contextualize these phenomena within broader spatial and temporal structures. The studies show that collective violence, at least if it is sustained over a certain period of time, aims at organization and therefore develops constitutive and integrative mechanisms.

This book was originally published as a special issue of *Civil Wars*.

Jutta Bakonyi, PhD, worked at universities of Hamburg and Magdeburg and the Max Planck Institute for Social Anthropology in Halle (Saale), Germany. Currently she is with the Civil Peace Service in Kenya.

Berit Bliesemann de Guevara, PhD, is senior researcher and lecturer at the Institute for International Relations, Helmut Schmidt University Hamburg, Germany.

December 2011: 246 x 174: 160pp
Hb: 978-0-415-69562-6
£80 / $125